FROM ANCIENT TO FUTURISTIC WORLD:

CHINA'S ADVANCED JOURNEY THROUGH INNOVATION, AI, AND INFLUENCE

Michael Loong

SCRIPTOR HOUSE
THE EPITOME OF GREATNESS

Scriptor House LLC

2810 N Church St Wilmington, Delaware, 19802

www.scriptorhouse.com

Phone: +1302-205-2043

Published by Scriptor House LLC

Paperback: 979-8-88692-047-5
eBook: 979-8-88692-048-2

CONTENTS

CHAPTER 1
THE NEW WORLD ORDER

During the founding of the People's Republic of China in 1949, China's industry was backward, capital and skill were lacking, and farmers' productivity was low. Many people were illiterate and unemployment was dire, affecting the country's social stability and security, especially in the urban environment. Then the central government set up many state-owned enterprises (SOE) offering people employment opportunities. Production efficiency was not an important consideration. It was better to offer employment to people rather than offering unemployment assistance. SOE does not need to project a profit margin for alleviating unemployment problems for the time being. Survival was paramount. Starvation is not an option.

After suffering from many years of failed economic policies and challenges during the national reconstruction period, Deng Xiaoping in 1978 introduced a new policy of market economy and finally succeeded in embarking on continuous positive social and economic development. The SOE had also improved with increasing production efficiency. In just five decades, successive Chinese leaders worked hard to serve the Chinese people, finally achieving the world's second-largest economy in 2012 and lifting 800 million people out of poverty. After China became a member of WTO on 11 December 2001, China exports began to accelerate. It has bilateral trade with over 150 countries and is the world's largest exporter. In 2022, China import was worth US$2.71 trillion and export US$ 3.57 trillion with a trade balance of US$1.86 trillion. In 2023, the US will have a trade deficit of US$336 billion with

China. The US consumers enjoyed many years of goods made in China with high quality and low price.

In the past 30 years, China has built the world's longest high-speed rail network of 45,000 kilometers across the country, the total length being more than the rest of the world combined. On January 3, 2019, China's Chang'e-4 probe successfully soft landed on the far side of the moon, allowing humans to see the dark part of the moon for the first time. Also, to probe the geological characteristics and origin of the moon. On November 24, 2020, Chang 'e-5 probe successfully drilled 1,731 grams of lunar soil and brought it back to Earth. China's BeiDou navigation accuracy is more accurate than the US GPS. It provides an accuracy of less than 1 meter, while GPS provides an accuracy of 4.9 meters. Driverless cars use BeiDou to navigate more safely than GPS. Huawei autonomous cars can self drive and park.

China manufacturing represents 35% of global manufacturing output, 80% of global cotton production, 50% of world iron and steel production, 30 million vehicles production in 2023 and 20% of the global medical product market. Most of the world's consumer electronic products are manufactured in Shenzhen, and most world commercial ships were built in China. China produces most solar panels and wind turbines in the world. Its integrated supply chain is the most comprehensive in the world. Shanghai has the largest container port in the world. China's trade surplus reaches $1 trillion in 2024. China is the world's factory. China is the first country to invent paper money and also the first to discard paper money by use of digital currency. China has the largest robotaxi fleet in the world. China's technological advance is on par with the US, and in certain areas more advanced than the US.

In order to help the remaining 600 million rural farmers out of poverty, China is to build 200 new smart cities that lead the way in adopting advanced technologies. This includes artificial intelligence, IoT, big data, cloud computing, transport, public security, the environment, healthcare, and manufacturing. It is an opportune time to build the smart cities, as China's economy is in a doldrum state presently. The investment in the smart city project will bring immense economic returns to the country. The government will provide skill training to all the rural residents who have moved to the new smart city. Eventually, they will enter the job market and enjoy higher income than before, thereby contributing towards the national economy.

China will establish a new state-of-the-art smart provincial E-government to minimize manual operation. It will reduce government expenditure, avoid human error, and improve the efficiency of the administration of the central government and provinces across the country. It will also monitor suspected terrorist activities, especially in Hong Kong, Xinjiang, and Tibet, as the US is determined to undermine China's security and incite internal uprising to topple the government. Regime change is the US favorite foreign policy against other countries.

In an era of a digital world, Chinese people are already using mobile phones to pay for goods and services and paper money will soon be obsolete. Soon, the digital currency of RMB will be used in cross-border trade between China and other countries . China is determined to be self-sufficient in food supply by establishing agricultural factories producing grains using artificial intelligence and IoT. The Chinese airlines have started using the newest China-made Comac C919 aircraft. With the visa free entry policy introduced in 2023 for many countries , tourism business has increased many folds, contributing more income towards national treasury. Free visa travel allows more foreigners to appreciate China's social and economic achievements that have improved Chinese people's living standard. Foreign visitors can experience how safe China's cities are for people to walk in the streets at night without fear of being harassed or life-threatening encounters.

In 2023, China's EV export sales exceeded that of Japanese cars, and China's BYD EV sales in China surpassed that of Tesla EV. It is amazing that BYD only started to make EV in 2009, and in 2023 it will become the world's largest EV manufacturer. Many Chinese EVs are exported worldwide. Chinese EV manufacturers are establishing production overseas in Spain, Poland, Hungary, Thailand, and Malaysia. Many foreign car manufacturers in China are having joint ventures with Chinese battery manufacturers to compete in the Chinese EV market. Many countries are using solar panels made in China, as they are cheaper and of better quality. China is embarking on a large scale of renewable energy projects to reduce the use of fossil fuel.

China imported $350 billion (more than 2023 Pakistan's GDP of $338b) worth of semiconductors in 2020—more than the value of crude oil being imported the same year, indicating the strength of Chinese electronic industries. In 2020, China had an international trade surplus of $570.4 billion. Now that the US imposes sanctions on chip exports, China has to manufacture in-house to compensate for the shortfall. The Chinese government now offers financial incentives of US$47.5 billion to local manufacturers to boost the semiconductor industries, compared with the US government's US$53 billion grant for chip manufacturing.

After the US bans the sale of chips to Huawei, Huawei stops selling smartphones for three years. In 2023, Huawei broke through the chip blockade and produced the new Huawei Mate 60 Pro 5G mobile phone with the 7 nanometer chip produced in-house. The phone has satellite communication ability, which no other phone makers possess. In an instant, the sales of Huawei mobile phones surpassed that of the iPhone in China. Chinese drones made by DJI in Shenzhen are selling like hot cakes worldwide, as the drones are also used in huge numbers by foreign countries.

BRICS now has nine members since January 2024 . Many more countries are now waiting to join BRICS. The BRICS is planning to implement a de-dollarization program in bilateral trade amongst members using their own national currency instead of the usual US dollar. Many countries have a fear of the US imposing sanctions on them and the US may confiscate their US dollar bank account. The US is using the dollar as a political weapon to threaten other countries. The US has frozen Russian and Afghanistan US dollars in US bank accounts. Now the US threatens to confiscate the interests accrued to Russian bank deposits to fund the Ukraine war operation. When the war occurred in Ukraine , the US and EU imposed sanctions on Russia oil and natural gas exports. Oil prices soared around the world, including in the US. The war in Ukraine affected Russian exports of fertilizers to other countries.

After the US and EU companies exited the Russia domestic market, the bilateral trade between China and Russia soared to new heights. Chinese EV dominated the Russian car market. At the BRICS Business Forum on October 17-18, 2024 in Moscow, the BRICS group of nations introduced a new payment system called BRICS Pay. Participants are given new cards loaded with 500 rubles, which they can use for purchases at designated shops marked with the BRICS Pay label. This initiative represents a significant move towards enhancing local currency transactions among the 159 countries poised to adopt BRICS

Pay. The system aims to reduce reliance on the US dollar and traditional financial networks such as SWIFT.

Since China attained the second-largest economy of the world title in 2012, it has become a significant player in the global political forum. Many countries including some European countries want to meet up with Chinese leaders to discuss economic, industrial and social cooperation. Many African leaders comment on the differences between the Chinese and American foreign policy. The American Secretary of State visits African countries only to talk about China's threat and debt trap whereas the Chinese foreign diplomat visiting African countries talks about building infrastructure like highways, railways, ports, hospitals, and dams. Many countries consider China as a friendly country willing to help without political interference. China's rapid economic success inspires many Third World countries to emulate it.

Today, China is the world's largest manufacturing and global trading country, as well as the world's largest market. In the past, people in many countries were fearful of visiting China for being unsafe. But now they are eager to visit China to attend conferences and seminars on a variety of topics concerning science and technology, social, environment, education, trades, culture, sports, arts, music, climate change, IT, AI, aviation and car exhibitions, machineries, robots etc. China is completely an open country, safe to travel and tour.

In 2023, China will be the top manufacturing country, with 31.6% of the total global manufacturing output. China accounted for 19% of the global economy in 2022 in PPP terms. In 2023, 70000 patents were granted to China, the highest in the world. In 2019, the number of outbound tourists from China reached nearly 155 million before the COVID-19 pandemic. China was the country with the highest outbound tourism expenditure worldwide in 2023, with Chinese travelers spending US$196.5 billion on outbound trips. This amount is larger than the Ukrainian GDP of US$188 billion.

China is the world's factory. It accounts for 50% of the world steel and aluminum production, 70% global production of rare earth elements, 80% of global solar panel production, the largest EV manufacturer and exporter in the world and the largest global chemical producer. China represents 60% of world cement output. In 2024, China produced 30 million cars, the most in the world. China exports more than 5 million vehicles in 2023, topping Japan to become the top country for car exports in the world. China's factory can produce an EV every 76 seconds. DJI controls 70% of the global consumer drone market. More mobile phones are made in China than other countries. Chinese factories can produce one smartphone every second. The US relies heavily on China for pharmaceutical products including active pharmaceutical ingredients and supplies of generic medications e.g., antibiotics, pain relievers, penicillin. In 2022, the US imported $10.3b worth of pharmaceutical products from China. China supplies 95% of ibuprofen, 91% of hydrocortisone, and 70% of acetaminophen to the US. China's supply chain ecosystem is unparalleled in other countries. China now exports more to the emergent markets than to North American and Western Europe combined. Hence, US sanctions or high tariffs on China do not do much harm to the Chinese economy, which is broad based. As China autonomous AI factories are gradually replacing human workers with robots, China can reduce legal working hours from 40 to 25 hours per week.

In January 2025 when Donald Trump became the 47th US president, he started a global trade war imposing heavy import tariffs , raising the ire of all the US allies and other countries. They reacted with similar tariffs on imports from the US. He is creating a crusade against the world by disrupting world trade and the global supply chain. China imposes tariffs on American agricultural products, making their products uncompetitive compared with similar agricultural products from Brazil and Argentina, who now become China's alternate source of agricultural product imports. Many American farmers have suffered bankruptcy. The same fate befalls many American agriculture equipment manufacturers, who also suffer from higher costs of imported steel with high import tariff.

China is the world's largest producer of rare earth producing 270000 tons in 2024. China imposed a ban on exports of rare earth minerals to the US. Rare earth minerals are used in many industries that make smartphones, military weapons, computers, TVs, magnets, camera lenses, EVs, wind turbines, batteries, nuclear reactors, medical equipment, lasers etc. China controls the global supply of rare earth minerals, accounting for 70% of the world's production and refining capacity. The scarce supply of the rare earth minerals will cause the US untold difficulties in manufacturing anything that requires rare earth minerals. Donald Trump finally was compelled to call President Xi Jinping seeking China to rescind the export ban of rare earth minerals to the US on June 9,2025.

As a result of Donald Trump's worldwide crusade policy of massive tariffs on imports, many countries are flogging to China to seek access to their products to Chinese markets. China's economy will boom as long as Trump is in power in the US. The US's loss is China's gain in world trade. Donald Trump's global tariff policy turns other countries to trade more with China, thereby reducing their trade with the US. It is a bonanza to China for many years to come. Irrespective of the

US continuous anti-China propaganda and military threat, China will continue to forge ahead with its economic policy coupled with its 2030-2035 long term economic 5-year plan in the following 10 projects: Advanced IT, advanced machine tools and robotics, Aerospace equipment, Maritime equipment and high-tech shipping, Modern rail transport equipment, NEV, Agriculture equipment, New materials and Biopharma & advanced medicinal products. These projects will be carried out by successive leaders.

History has proven that the Chinese people have survived 5000 years through wars and peace and 200 years of humiliation of Western power invasion. There are 1.4 billion Chinese people with unity, creative minds, resilience against external threats and latest high-tech weaponries. Hence, it is an exercise in futility for the US to crush China's survival by sanctions, trade war , chip war, tech war, tariff war, war of any kind, CIA propaganda, and the naval ships in the South China Sea, wanting to fight a high-tech war with China halfway around the war and expect to win is a pipe dream. It is a leap of faith to cripple the economy of China, which trades with 150 countries around the world. It defies logic that Americans can live a good life indefinitely by continuously borrowing money from other countries. The world can survive without the USA, but the USA cannot survive without the world. The US government is too busy dealing with proxy wars around the world to serve the pockets of the MIC and Wall Street.

1.1 THE EARLY 21ST CENTURY BELONGS TO USA

America's power in the 21st century depends on five elements: US dollar, military weapons, technology, finance, and petrodollar

- US Dollar

The dollar has become the currency for world trade and reserve currency and is no longer pegged to gold. Every country has confidence in the dollar and it has a stable value. When traveling abroad, the dollar can be exchanged for local currency easily. International commodity price is based on the dollar. The Fed is ready to print unlimited amounts of fiat money for use around the world. The Fed has the power to raise or lower the interest rate on US Treasury bonds at any time, thus affecting the interest rates of other countries around the world. The high interest rate affects every country's economy and cost of living. The US government has threatened many countries with freezing their accounts in US banks if the US finds them unfriendly. The Russian government's US dollar accounts in the US and the EU have been frozen because of the war in Ukraine.

- Military Weapons

The US has more nuclear weapons, fighter jets, aircraft carriers, warships, ballistic missiles, tanks, and all kinds of ground armaments than any other country. The US has enough weapons to wipe out every country in the world. The US has 750 military bases in 85 countries and territories. The military industrial complex (MIC) that makes weapons is fully sponsored by politicians in the Congress. The US military budget increases every year. Military contractors charge the government on a cost-plus-profit basis, and as a result, the shares of American weapon manufacturers are cash cows to politicians. The US foreign policy is to create fear amongst every country by labeling China, Russia, Iran, and North Korea as potential (invisible, fake) perpetual military aggressor and hence they should buy US weapons to safeguard their security. Once any country buys US weapons, this country is forever locked in by US control as US weapons operation is coded and administered by US military personnel.

In 2025 when Donald Trump foreign policy upset EU countries , German officials were sounding the alarm over what they considered as a" kill switch" embedded in the US made F-35 fighter jet. The switch could be used to deactivate or limit the combat function of the F-35. The Portuguese government intends to buy other fighter planes instead of the F-35. In September 2024, thousands of handheld pagers intended for use by Hezbollah exploded simultaneously killing at least 42 people and injuring 4000 civilians. The pagers have been secretly embedded with explosives and activated remotely by an Israeli intelligence agency. In Nov. 2024 Israelis Prime Minister Netanyahu admitted responsibility for the explosion.

- Technology

American technology is superior to other countries in the fields of electronics, aviation, telecommunications, chemistry, IT, medical, and military equipment. The US attracts many scientific research and graduate students from many other countries, and the immigration policy of the US actively attracts foreign talents to immigrate to the US. There are more Nobel Prizes given to American scientists than other countries. Many Chinese science students remain in the US after graduation to do research and work in high-tech companies.

- Finance

The US financial system controls global financial operations. Its credit cards can be used worldwide. In 2008, the outbreak of the subprime mortgage financial crisis in the US affected the global financial market and triggered the global economic depression. Wall Street financial tycoons not only control investments in the US economy, but also influence investment in other countries. Stock prices on the New York Stock Exchange can be manipulated up and down by them. When stock prices go up and down, they profit in either direction. The Wall Street stock market is unpredictable, as the stock prices can go up even during an economic downturn. Periodic Fed rate hikes have led to global financial crises and foreign currency devaluations in 1997, 2008 and 2023.

When the Fed interest rate is low, hot money from the US flows into the capital markets in various countries seeking borrowers. When the Fed raises interest rates, hot money flows back to the Fed, causing foreign currencies to depreciate. Every international trade in US dollars must go through the SWIFT system to effect the fund transfer in dollars. The US financial institutes earn handsome commission from the two-way foreign exchange transactions. Their commission could reach hundreds of billions of dollars each year. When the BRICS members begin to de-dollarize in bilateral trade using local currencies, the US will lose substantial foreign exchange commission. The amount of USD circulation in the world will reduce.

- Petrodollar

In 1973, the petrodollar system was created by the US with Saudi Arabia, who agreed to price and trade oil in US dollars. Other OPEC countries followed suit. These OPEC countries will receive dollars in large amounts from oil sales, and they most likely use their surplus dollars to buy the US treasury bonds. Hence, the petrodollar will flow back to the US domestic financial market. These oil export countries will hold a large amount of USD issued by the US Fed. In 2022, the OPEC countries oil sale reached USD 888 billion in petrodollar. The oil importing countries need to buy US dollars to pay for the oil. The payment will go through the SWIFT system for settlement. The US financial institutes will process the foreign currency transaction into the dollar, and they will charge a commission. The US$888 billion petrodollar transaction could yield USD 8.88 billion revenue through double foreign exchange to the US financial institutes. The petrodollar is a cash cow. On June 8, 2024, Saudi Arabia and the US petrodollar agreement expired.

1.2 CHINA ECONOMIC REFORM

In 1978, China opened up the market to allow foreign companies to set up factories in China with tax incentives. Initially, the factories produce simple consumer products for sale to the world. The skill required to produce these goods is relatively low. Hence, the worker's salary is low compared with that in western countries. With the economy gathering speed, China begins to invest in industries with higher value on the value chain. China starts to export commercial ships, TV, washing machines, air-conditioning units, computers, smartphones, industrial tools and machineries, port cranes, EV, drones, EV batteries, solar panels, wind turbines, trains etc. The Chinese government has a long term plan for every 5-year period.

China is the top trading partner with over 120 countries. China is the world's largest economy since 2016 when measured by purchasing power parity (PPP). China accounted for 19% of the global economy in 2022 in PPP terms. China sets up the Asian Infrastructure Investment Bank (AIIB), Shanghai Cooperation Organization (SCO), BRICS, Regional Comprehensive Economic Partnership (RCEP) and Belt and Road Initiative. China's de-dollarization will reduce the cost of doing cross-border trade with other countries, saving China billions of USD annually in foreign exchange transaction costs through the SWIFT system. China has built the world's longest high speed rail of 45000 km., sending a space station into space and a capsule to the moon, WeChat and Alipay are used everyday by people to pay for purchases of goods and services. Farmers are using live-streaming to sell their products online.

The Chinese army uses drones to deliver cooked foods to soldiers stationed in remote mountainous posts. China is to start building 200 new smart cities to accommodate the remaining 600 million villagers and retrain them with new skills to work in industries. Once they become productive in high value manufacturing industries, China's GDP will surpass that of the US as 600 million of them is nearly twice the population of the USA. China's economic development has been fueled in large part by a sprawling industrial sector, which includes manufacturing, construction, mining, and utilities. In 2022, value-added industrial output accounted for nearly 40 percent of China's GDP, compared with that of the US 18 percent.

1.3 US PRESIDENTIAL ELECTION SYSTEM IS NOT DEMOCRATIC

The President of the US is elected indirectly by the Electoral College in December of the election year, rather than directly by the electorates. After the election polls are closed, the Electoral College votes in each state to determine the outcome of the presidential election. Each state has as many electoral votes as it has members of Congress. There are 538 electors in the US, and the candidate who wins more than half of the electoral votes is elected as the president. Members of the Electoral College gave Trump more than 270 electoral votes on December 19, 2016, officially electing him as the 45th president of the US, although his political rival Hillary Clinton won nearly 3 million more popular votes than Trump. It is not a democratic concept that the president of the US is elected by a small minority of voters, and not by the majority of voters.

The US does not have much democracy when it comes to choosing a president because there are only two candidates to choose from the two political parties. The candidate for the presidential election is selected by the party members. The US President holds the dictatorial power to decide for the country's future and foreign policy without congressional approval. President Donald Trump's 2025 global tariff is not approved by the Congress. President Joe Biden in December 2024 gave his son Hunter unconditional pardon when he was charged with two crimes.

Unlike other democratic countries where the people can hold demonstrations or protests to change the government, but not in the US unless the President is impeached by the Congress which may be controlled by the President's political party. In contrast to other democracies, the party which wins the majority electorate votes is elected as the government with a clear manifesto of what the party will do for the people. In the US, a presidential candidate will tell voters his policies on the campaign trail, but not everyone in his party agrees with his policies. For example, Trump wants to raise import tariffs on all Chinese imports, but not all members of the Republican Party agree because his policies would increase manufacturing costs for some businesses owned by some Republican members.

1.4 Relationship Between American Politicians and Rich Tycoons

The American economic system is centered on the Wall Street stock market and focuses on maximizing shareholder returns. At the heart of America's economy is 500 companies headquartered in the US, most of them in New York. Their manufacturing, purchasing, and sales operations spread over the world. These big businesses have no particular loyalty to the country, as most of their factories are located in foreign countries. The corporate executives only care for their shareholders to maintain their stock at a high price so that they can receive high salaries and bonuses. To avoid paying very high income tax in the US, many big businesses have their company headquarters registered in the British Virgin Islands (BVI) which is a tax-free jurisdiction for international business companies (IBCs). IBCs that are incorporated in the BVI but operate outside the country are not subject to any corporate taxes, including income tax, capital gains tax, and inheritance tax

In the pursuit of wealth, executives take various measures to increase the company's stock price, including reducing wages, cracking down on unions, and outsourcing to factories around the world to make cheaper parts. Boards of directors dole out sky-high salaries and bonuses to the company's top executives. Wall Street titans and US politicians have close ties. The titans provide donations to the politicians during their election campaigns. After the elections, some businessmen can get plump jobs in the government, like an ambassadorial position in a foreign country or a position in the treasury department. The retired politicians can become lobbyists for the Fortune 500 companies or join them as members of the board of directors.

The US Fed can print money freely and sell treasury bonds whenever the treasury is short of cash. During a financial crisis, the government can lend money to companies and banks that are too big to fail. This is a form of socialism. The US government subsidies the farmers and also companies that can produce chips that are not available in the domestic market. Every year, the US government has a deficit budget. The US government can raise the debt limit (or deficit ceiling) to meet its expenditure obligations, allowing the maximum amount it can borrow to finance its existing legal obligations, including Social Security, Medicare, and military salaries. If the debt limit is not approved by the Congress, the government will default on payment to federal government employees.

1.5 US Foreign Policy Towards China

The most exclusive foreign policy of the US towards China is to create domestic disturbance in Hong Kong, Taiwan, Xinjiang, and Tibet. The US CIA creates a worldwide propaganda to smear China with no human rights, no freedom of speech, suppression of dissidents. The CIA will pay any privately owned media to spread anti-China rhetoric with disinformation and fake stories. The CIA propaganda spreads disinformation in Africa, warning African countries China's investment is a debt trap. The anti-China propaganda is doubled down by the media in the West through proxy politicians in the EU parliament and certain right wing politicians in European countries like UK, Germany, Italy, Lithuania. The US smearing campaign against China will perpetuate for as long as China is considered a 'security' risk by the US politicians.

It is reported that the CIA has a department concentrating on creating disinformation and propaganda against China, and it has a budget of $1.6 billion in 2024. The ICA would pay a writer to write bad reports and fake stories about the Chinese government to be published by private news media. It also collaborates with Western NGO and news media to spread the bad reports and fake stories in the West. The CIA financially supports the Western NGO. In addition to smearing campaigns and propaganda against China, the US also has created AUKUS, a military alliance consisting of Australia, UK, and USA to contain and threaten China's security.

Australian Strategic Policy Institute(ASPI) describes itself as an independent, non-partisan think tank that produces expert and timely advice for Australia's strategic and defense leaders. However, ASPI elects to be an advocate for the USA to become a partner of US foreign policy. It alleged genocide of Uyghurs in Xinjiang. It also ran with the same US conspiracy theory that the Chinese government was responsible for the COVID-19 outbreak in Wuhan. It is being used as a conduit for hardline US policy towards China. According to report, ASPI is sponsored by the following foreign organizations:

- US State Department $762,559
- US Department of Defense $201,136
- US Embassy Canberra $400,576
- Lockheed Martin $25,000
- Northrop Grumman $67,500
- Rafale $20,000
- Raytheon $19,090
- SAAB Australia $25,000
- Thales Australia $113,334

ASPI recommends the Australian government to buy US nuclear submarines to be delivered in 2040 and cancels the original cheaper submarine contract with France. The contract of the US nuclear submarine is actually meant to help the US to boost its naval force in the South China Sea to threaten China. In 2025 President Trump issued an order to stop funding overseas programs which include funding for ASPI.

1.6 China's Strategy to Deal With US Threat

As long as there is no internal unrest in China and no threat of foreign invasion, China has a chance to advance its economy and to become the largest economy in the world in near future. The only real external threat comes from the US, but a hot war is unlikely between the two nations because China has nuclear weapons and a long range ICBM that can reach American land mass. If a hot war were to start by the US against China, Russia would join forces with China to fight the US as nuclear fallout will affect Russia. The US has formed military alliances with other countries and conducts regular naval exercises in the South China Sea. These provocative exercises have been around for many years and have not affected China's economy.

US warships have patrolled the South China Sea for more than 70 years as a show to Asian countries that the US is protecting their security against Chinese attack. Therefore, they should join the US military alliance and purchase US weapons. The usual American tactic to sell weapons to other nations is to create chaos and geopolitical tension in targeted regions. When the tension becomes as serious as the CIA propaganda depicts, the US Secretary of State and his companion Secretary of Defense will visit the worried nations to sell weapons for their defense requirements. Many of the weapons are embedded with chips that are controlled by the US in case the buyer nation turns anti-USA. The US can deactivate the weapons by long distance signal. Once a nation buys the American weapons, the US has reason to station military personnel in the host country, seemingly to train the local military personnel. Subsequently, the US can influence the host country's foreign policy.

China adopts historical experience in handling international relations. Confucius advocates using diplomatic means to resolve conflicts between countries first, to avoid imposing suffering on people caused by a war. War can destroy the country's economy, even if the war is won. The best foreign policy is for the country to work for peace and stability. It is expected the US will continue to cause unstable relations with China when China's economy surpasses that of the US. The US treats China the same way as they treat Russia as an imaginary enemy. The US started a Cold War with Russia for more than 70 years without starting a hot war, which the US tries hard to avoid. The US uses the Russian threat as a means to sell weapons to European countries. Similarly, the US treats Iran as an enemy to the US and other countries in the Middle East, so the US can sell weapons to the countries threatened by Iran, which is a Shia country. So the US will promote the sale of weapons to countries like Saudi Arabia and Turkey that are Sunni.

Although the US has been creating tension in the South China Sea and spreading anti-China propaganda around the world, China remains steadfast and firm using diplomacy and not using force to deal with the US. China needs to be patient as the US power is beginning to wane on the world stage. There are a plethora of domestic problems that could erupt into violent protests between the rich and the poor in the US. Some pundits predict a civil war could break out in the US due to Donald Trump's swift dismissal of many Federal employees conducted by DOGE controlled by Elon Musk in 2025. Time is on China's side to await the final curtain to fall on the US to stop its anti-China and military threat.

The US creates a lot of smoke screens of China threat around the world. China adopts flexible diplomacy to deal with each and every

American anti-China threat. There is no consistency in American foreign policy, and it changes with the change of political party President. Its policy tends to be disruptive in global politics and supply chain without the slightest care of repercussion. The US frequent use of sanctions to threaten other countries have caused many countries to detach from the US diplomatic relationship. Hence, many countries become more friendly towards China. In 2025, Donald Trump applies heavy import tariffs across all countries, raising the ire of US allies, especially Canada and Mexico.

The USA is a country full of surprises, contrast and always in a flux. It is a country with mesmerizing charm and intrigue, cunning and skill with an exceptional corruptive behavior. Even though the US military operations in Vietnam and Afghanistan had killed many innocent people, yet many Vietnamese and Afghans refugees were keen to migrate to the USA. Many Chinese smart students go to the US to study and after graduation prefer to remain in the US working in the high-tech industries which manufacture weapons threatening the security of China. President George W. Bush said openly to all nations: "you are either with me or against me". At the same time, Secretary of State Henry Kissinger said to the world:" To be a friend of the US is fatal". Hence, either way, no country can avoid being an enemy of the US. So once you are in the cross-hair of being an enemy of the US, you are unlucky. Saddam Hussein, Osama bin Laden and Bashar al-Assad were US "enemies" although Saddam Hussein was once a US ally.

The US is well-known for its foreign policy of regime change of foreign countries. China foreign policy consistently avoids getting involved with American dirty politics around the world. The US as a country is very complex and contradictory. It declares itself as a protector of human rights and democracy. Yet its warmongering history shows a contradictory, different story. The country is controlled by two powerful entities: BlackRock and MIC, which together control the governance of the country. The US has a vast deep state that controls the political agenda of the country. The powerful deep state groups could enlist a 20-year-old assassin named Thomas Matthew Crooks in an attempt to assassinate Donald Trump during his campaign speech at Bethel Park, Pennsylvania on July 13, 2024. Crook had no prior criminal record. A similar assassination attempt on Ronald Reagan happened in 1981.

The most sinister assassination happened to President John Kennedy, who was shot dead on 22 November 1963. Senator Robert Kennedy, a Democratic presidential candidate and a brother of J Kennedy, was shot during his campaign speech at the Ambassador Hotel in Los Angeles and died on June 6, 1968. Three other US presidents had been assassinated, and they were Abraham Lincoln in 1865, James Garfield in 1881 and William McKinley in 1963. History of the USA shows many political leaders were targets of assassination starting from the early 19th century. Political assassination culture in the US is no different from some underdeveloped countries. Personal freedom overrides other human values. Make America great again is for the rich and not for the masses.

The US has a nickname of being called a cowboy country. The slogan "American Dream" has a powerful mystical call to all people in other countries to go to the US to seek the dream, even if they can end up as a victim of hate or gun violence crime. In The US, it is well known that people do not walk in the streets at night for fear of being harassed or attacked. US President Donald Trump has a supreme power of issuing an executive order on USAID to be disbanded without Congressional approval. US President Joe Biden can issue a pardon for his son

Hunter Biden sentenced for criminal misdemeanor. This is how the US politicians define democracy and human rights.

The USA was a colony of Britain for about 167 years before independence. Since independence 1776. The US has been involved in 115 military conflicts. The US has the highest rate of civilian-owned firearms in the world, with about 120 guns for every 100 Americans. Nearly 43,000 people died from gun violence in 2023. As of the end of 2022, the prison population in the US was 1,230,100 people. The US leads the world in total number of people incarcerated. On 25 May 2022 19 school children, including two 10-year-olds, were killed by an 18-year-old gunman in south Texas. Even some Chinese students were murdered in the US. The US is a violent and not a peaceful country⊠Such is the American personal freedom and human rights.

In 2023, there were more than 277000 Chinese students studying in the US. Their parents were not deterred by the gun violence in the US. Secretary of State Henry Kissinger once said to the world: "To be a US enemy is dangerous, to be a US friend is fatal". His statement is clearly epitomized by the US 3-year-old proxy war in Ukraine, fighting for the USA to cripple the Russian economy and to topple President Putin's government. There is no winning chance for Ukraine to engage in a war with Russia, which is a large country with more soldiers, resources, and weapons than Ukraine. Ukrainians did not learn a lesson from history that a country as strong as Germany or France failed to win a war against Russia in 1941 and 1812 respectively. Even the most powerful country USA in the world cannot win a war in Vietnam and Afghanistan fighting the poorly equipped soldiers. Ukraine President Zelenskyy wants to be a friend of the US.

1.7 TRADE RELATIONS BETWEEN CHINA, RUSSIA, AND THE EU

China and Russia relations have taken a turn for the better after the war in Ukraine started in 2022 and the heavy US sanctions against Russia. Russia's trade was immediately severely curtailed when the US and EU companies all withdrew from the Russian market, resulting in trade between China and Russia soaring. Sales of Chinese electric cars are growing rapidly in Russia. Many Russian supermarkets sell Chinese products. The use of Chinese RMB by traders has increased many folds. China also imports more oil from Russia than before, as Russian oil is cheaper than other suppliers. Bilateral trade between Russia and China in 2024 reached US$ 240 billion, compared with US$147 billion in 2021. The US proxy war in Ukraine has forced China and Russia relations to be more consolidated. Traditional US foreign policy is to drive a wedge between China and Russia so that the US can divide and rule. The unpredictability of US politics is unique in geopolitics.

Germany's booming industries have long depended on cheap Russian natural gas. Since the damage to the Russian Nord Stream gas pipeline linked to Germany caused by the US, the German economy suffered badly without the cheap gas from Russia. It needed to import higher priced natural gas from other countries, including the US. Such problems had caused German industry production costs to increase, thereby reducing its competitiveness. German export industries had suffered significantly. In 2024, Germany was in recession. Such a situation is bad for the EU, as Germany is the largest economy in the EU. Some German companies have shifted their production overseas. Meanwhile, Germany invests more money in China in 2024 to boost its factory outputs. Meanwhile, despite economic doldrums, Germany continues to increase its military and financial assistance to Ukraine to fight Russia. Proxy war has become a fashionable game of international geopolitics.

The EU has been China's second-largest trading partner after ASEAN. Its economy is beginning to suffer as the US demands the EU to decouple from China. Following US demand, the EU banned the use of Huawei 5G communication equipment and raised the import tariff on Chinese EV imports. Dutch ASML also stopped selling EUV semiconductor equipment to China. The US pressures the EU to ban the import of cotton from Xinjiang. There are many EU politicians who have anti-China sentiments. China is a huge market for EU products and if the EU continues with a hostile attitude towards China, the EU would suffer economically.

The EU and UK have now changed their foreign policy towards China because of the heavy tariffs imposed on goods imported to the US by Donald Trump in 2025. They begin to ask China for a better bilateral trade relationship. With the Ukraine war still in progress, the economy of the EU and UK are in doldrums. These countries have been heavily supporting Ukraine financially and militarily, draining their treasuries. Chinese EV and battery manufacturers are setting up factories in some European countries. In 2024, China-EU bilateral trade reached $762 billion, a 1.6% increase from the previous year. In May 2025, the EU sent a delegate of high officials to visit China to discuss bilateral trade when the EU was threatened by the US high tariff.

1.8 FUTURE RELATIONS BETWEEN CHINA AND JAPAN

China and Japan are neighbors. Both have a long historical relationship. Japan became technically advanced when it adopted Western culture and modernization during the Meiji Era in 1868. Japan began to invade China in 1931. In 1945, Japan was defeated and surrendered to the US during WW II. Since then, the US has a military base in Japan to control Japanese diplomacy and military operations. Japan became the second-largest economy in the world in 1968. Trade between Japan and China flourished after China opened up its market. By 2007, Sino-Japanese bilateral trade reached US$236.6 billion, about 17.7% of total Japanese external trade, exceeding the trade volume between Japan and the US. In 2023, China-Japan trade grew to $266.4 billion.

Japan is an island country, with an aging population and little arable land for agriculture. It suffers from frequent volcanic eruptions and earthquakes. The problem of survival is a worry to the Japanese people. It must do international trade to have the economic ability to maintain its standard of living. China's huge market is good for Japan's trade. The quality of Japanese products is favored by many Chinese consumers. China's industry in some areas has not yet reached the level of Japan's international quality and after-sales service standard. China still needs to import precision instruments, machineries, and tools from Japan to manufacture many quality products.

China and Japan will remain in a friendly relationship for a long time, avoiding any military conflict. However, presently, Japan may need to take sides amidst the geopolitical tension between the US and China.

Japan relies on exports to survive, and its ships have to pass through the South China Sea. If there is a military conflict with Japan, China can block Japanese ships in the South China Sea. In 2025, Donald Trump's high tariff on Japanese imports does not bore well for Japan-USA trade relationship. Japanese people need to understand China and Japan are neighbors forever and the distant US cannot protect Japan forever when Japan is of no more use to the US interest.

1.9 BELT AND ROAD INITIATIVE

The Belt and Road Initiative (BRI) was launched in September 2013 by China. There are 155 countries and 32 international organizations joining the BRI. African countries are urgently in need of infrastructures. In 2020, only 43 percent of Africans had access to electricity and 48 percent had access to paved roads. So far China has built seaports, railway, highway, power generating plant, schools, water supply and sanitation in the following African countries: Mozambique, Algeria, Tanzania, Uganda, Djibouti, Egypt, Ethiopia, Kenya, Nigeria, and Sudan. Under the BRI, China's reputation in Africa surpassed that of the US, as the US had not built any highway or railway for African countries. China's reputation in South America also gains momentum by construction of dams and ports. In Saudi Arabia, China built Mecca Metro at a cost of $16.5b.

In 2016, China built Indonesia's first high-speed rail for the 140 km line between Jakarta and Bandung. The high speed rail shortens the journey time from over three hours to forty minutes. The project was completed in August 2023. The line was officially opened on 2 October 2023. So far, the Jakarta-Bandung HSR carries 1.45m passengers in the first 100 days. The BRI flagship project in Laos is the Boten–Vientiane railway line running between the Laos capital, Vientiane, and the northern town of Boten on the border with Yunnan, China. The line was officially opened on 3 December 2021. It is China's first overseas railway project that connects to China's railway network. It will be Laos' longest line, connecting with Thailand to become part of the proposed Kunming–Singapore railway, extending from the Chinese city of Kunming and running through Thailand and Laos to terminate at Singapore. The China-Laos Railway is used for transporting produce from Thailand and Laos that is then sold in Sichuan and Guangdong in China. In addition, China's agricultural and industrial products are transported to Southeast Asian countries via the rail link,

1.10 CHINA ECONOMY AFFECTED BY US SUBLIME MORTGAGE CRISIS

The US subprime mortgage crisis occurred in 2008 due to the housing bubble burst caused by cheap credit and relaxed lending requirements. Subprime mortgages were taken on by many low-income households on expectations of housing price ever rising, and then they sold the mortgaged house to make a capital gain. When the interest rate began to rise rapidly, many people were unable to repay the housing loan. Prices of houses began to fall, resulting in a housing bubble burst. Lehman Brothers had many mortgage-backed securities in the housing market. It filed for bankruptcy on September 15, 2008, with $619 billion in debt. Its bankruptcy triggered a global financial crisis. Many banks and private investors around the world incurred huge financial losses. The financial crisis caused China a drop in the year-on-year growth rate of its GDP, falling from 13 percent in 2007 to 6 percent by the fourth quarter of 2008. The government quickly adopted a number of stimulus policies to avoid further slowdown of the economy. As a whole, China's exports fell by about 17% in 2009 before recovering to positive growth in 2010.

The housing bubble seems to be a game played by the US financial market tycoons, who manipulate the housing market like they do to the stock market. Manipulating the prices of stocks going up gradually and then suddenly crashing the prices down. The tycoons will scoop up the low price stocks and then gradually manipulate the prices to go up. Tycoons have immense wealth to move stock or housing prices up and down (within legal regulation). China helped the US government to avert a financial meltdown crisis due to the subprime mortgage crisis by buying more treasury bonds to help the US government with additional liquidity to bail out those companies too big to fail. More money is given to the banks for lending and preventing Freddie Mac and Fannie Mae from closure, and increasing liquidity for the Fed quantity easing policy to avert a severe recession. China did not receive a "Thank You" card from the US government. The capitalistic US government also practices socialism with American characteristics.

1.11 INTERNATIONALIZATION OF CHINESE RMB

China established the Asian Infrastructure Investment Bank (AIIB) in 2015. It is a financial entity that allows China to promote the digital yuan as an international currency that can be used directly in international trade transactions, avoiding the use of the SWIFT, which the US uses to sanction other countries. Expanding China's digital currency transactions with many other countries will strengthen the Chinese yuan's position as a global trading currency. Diversified international trade will reduce the risk of trade war with the US. The internationalization of RMB will reduce the US dollar in the global financial market. China can offer loans in US currency to the other countries and the repayments in RMB, thereby encouraging the borrowers to sell goods to China to settle transactions in RMB. The US frequently uses the US dollar as a political tool to sanction other countries. China also needs to increase its efforts to improve and open up its financial market. Presently, Chinese RMB is not freely convertible to the other currencies.

1.12 Future Relationship Between China and the US

The US and China are the world's two largest economies. China is now the third-largest export market for the US, behind Canada and Mexico. In a 2019 study, two American economists found that increased trade with China boosted the annual purchasing power of the average US household by $1,500 between 2000 and 2007. Bilateral trade is a win-win situation. A report by the US-China Business Council found that exports to China supported more than one million jobs in the US. American companies in China earn hundreds of billions of dollars annually from their operations in China. US bilateral trade with China totaled $758.4 billion in 2022. Exports were $195.5 billion and imports were $562.9 billion. China annually has a trade surplus with the US. In 2016, China overtook the US to become the world's largest economy in terms of purchasing power parity. According to the American PwC report, China is projected to be the world's largest economy, contributing about 20% of the world GDP in the near future.

Since Donald Trump started a trade war with China in 2018, the relationship between the USA and China began to get fractured. Donald Trump set high tariffs on Chinese imports and trade barriers on Huawei 5G equipment, and banned American exports of chips to China. The US continues to sell weapons to Taiwan and bribe young Hong Kong activists to demonstrate with violence against the Hong Kong Government. Then the US accused China of committing genocide in Xinjiang and hard labor against the Uyghur people. Such propaganda is to smear China's image around the world. In 2024, the CIA has a budget of US$1.6 b to carry out anti-China propaganda around the world. The US rallied its allies in the EU to decouple from China.

Joe Biden imposed a series of export control measures in 2022 with the aim of restricting China's access to advanced AI chips made with US technology in order to prevent China from improving its military modernization. In 2024, Joe Biden imposed a tariff of 100% on Chinese imports of EV. The US continues to involve its military allies the UK and Australia to join its naval exercise in the South China Sea. The US trade war against China will continue for a long time. Despite all the threat of military confrontation and creating friction between China and its neighbors, it is not envisaged the US would start a hot war with China. The US participated in 10 war games, and it did not win a single war game against China. The US fighting a prolonged war of 20 years cannot defeat the poorly equipped Taliban foot soldiers, what chance could the US have to defeat China which has nuclear weapons and hypersonic missiles fighting at the door steps of China.

Since China allows visa free travel for 15 or more days to many foreign visitors around the world starting from 2023, tourism has promoted the soft image of China. Many first time tourists from the US , Europe and other countries are shocked to see the modern structures of high rise buildings, convenient and cheap public transport by subway which is clean and on-time arrivals. The subway stations are huge and of modernistic design, looking more like an international airport. There are share-bikes in the cities. The streets are not as noisy as expected due to many EV on the roads. Many tourists are amazed that many public parks are filled with people strolling , exercising and dancing at night. Many tourists feel safe even walking in the streets well past midnight. The use of mobile phones for payments of goods and services make the tourists feel really convenient. There are no beggars, pickpockets or homeless people in the streets. The varieties of food make the tourists feel gastronomically delighted. The 350 km/hour high speed train ride gives all the tourists a big surprise of comfortable and smooth ride at an affordable price.

Some American tourists went to Xinjiang to visit and found the local people living happily and the cities full of vibrancy and prosperity. Their experiences debunk the American alleged hard labor and genocide in Xinjiang, where cotton harvests are done by autonomous harvesting machines . Some tourists feel that living in China is like living in a futuristic country. Visa free tourism has helped promote the soft image of China greatly. After the US banned the download of TikTok, which had been used by over one hundred million Americans, users switched to another Chinese app called RedNote. Through RedNote, they discovered that much of the negative information spread about China in the US was false. They discovered that Chinese people pay lower prices for groceries, have affordable healthcare, no annual property tax, and low university fees. Once mortgages are paid, homeowners fully own property without further taxes.

1.13 Signs of US Power Waning

The storming of the US Capitol Hill by Trump supporters on January 6, 2021, may be remembered as a turning point in American history. The revolt, fomented by Trump himself, raises deep questions about what kind of political system future American generations will inherit. In addition to polarizing the country to an incredible degree, Trump's single term of administration (2017-2021) has fundamentally damaged American institutions and political norms as well as global supply chain permanently. During his second term of presidency beginning in 2025, he doubled down on his policy of increasing tariffs on steel and aluminum imports from all countries including his allies, and a 145% additional tariff on Chinese imports. These tariffs will only increase the prices of imported goods from China, and the American public will need to pay more from now onwards. This will cause inflation in the US. The price of cars will increase significantly as every family needs a car.

China also announces a new series of retaliating measures against the US, including 15% tariffs on chicken, wheat, corn, and 10% tariffs on several agricultural products. China bans the exports of rare-earth minerals, including gallium, germanium, antimony, and super hard materials. These minerals are required by MIC weapon manufacturers in the US, and China is the major supplier of these minerals. In June 2015 Donald Trump made a call to Xi Jinping, begging him to release the sale of rare earth minerals to the US.

Ship building in the US is in a dire strait as the industry can build five ships in a year, whereas China can build 300–400 commercial ships and 10–15 naval vessels annually. The US Navy estimates China's shipbuilding capacity is 232 times larger than the US. US DoD reveals that 40% of semiconductors that sustain DoD weapons systems and associated infrastructures are sourced from China. The Pentagon imports Chinese DJI drones despite being banned by the Congress.

The Ukraine war is splitting the alliance of NATO members. Trump is a fervent peace advocate compared with Joe Biden who was the instigator to start the war. Trump also wanted the European members to increase their contribution to NATO operations to 5% and threatened the US withdrawing from NATO if his demand was not met. Trump also set up a new department of government efficiency DOGE to reduce waste and reduction of federal expenditure. Many thousands of Federal employees have been fired from their job. Billionaire Elon Musk was tasked to implement the drive to cut costs. He found many social welfare payments have been given to people who are more than 150 years old. The Pentagon fails many audits on its 61% of its $3.5 trillion in assets which are missing. DOGE wants to investigate many Congressional members having more than a few hundreds of millions in their bank account when they draw only a few hundreds of thousands dollars annually from their salary.

Tether cryptocurrency is used widely instead of Swift to transact international trade. It has become an important currency in the global financial system, About $190 billion changing hands every day. Tether is a digital US dollar (USDT). One USDT is equal to $1.00. Tether is unregulated and cannot be used as a political tool by any country. Tether market cap is $143.44 billion. Tether holding company in 2024 earned $1.3billion in profit during the second quarter. Russia is using USDT to do oil trading with other countries, bypassing US sanctions. It is convenient to use and faster in operation As Tether digital dollar operates outside US banking systems, countries like North Korea can do cross border trade using USDT. Asian countries with significant trade remittance flows could use USDT tokens to settle global payments in real time, reducing friction and costs. The US would lose substantial revenue from double foreign exchange fees in the Swift system, which is discarded by Tether uses.

USDT is a cryptocurrency stablecoin launched in 2014 and is distinct from a central bank digital currency (CBGC). USDT stablecoins are blockchain-based currency. It is owned by iFinex, a Hong Kong-based company that also operates the Bitfinex cryptocurrency exchange. The company that issues USDT is Tether Limited, a subsidiary of Tether Holdings Limited, which is registered in the British Virgin Islands. As of July 2024, Tether has more than 350 million users worldwide and holds more than $97.6 billion in US Treasuries.

Donald Trump in 2025 demands Ukraine to pay back all the financial and military aids amounting to US$350billion as claimed by him. Such demands shock the EU leaders. They wonder if there is a war in the future, would the US help the EU and then demand for payment of aid. Some NATO members are skeptical about the US possessing a "kill switch" to immobilize their F36 jet fighters if and when there is a disagreement with the US on foreign policy. Portugal is on the verge of cancelling F35 orders. Canada is also pondering on rejecting the F35, as Donald Trump wants to turn Canada into the US 51st State after imposing heavy tariffs on Canadian imports. The relationship between the USA and its allies are at the lowest ebb.

The three-year Ukraine War and the resulting conflict between Donald Trump and EU leaders on how to resolve it have left the latter unhappy, as Donald Trump has been secretly helping Russia devise a ceasefire deal in favor of the aggressor. The deal was planned without the participation of both EU and Ukraine leaders. The EU and UK prefer the war to continue with more aid to Ukraine, but Donald Trump wants the war to stop immediately. The EU is apprehensive that in the future , the US is unwilling to fight for EU security against Russia. In March 2025, the European Commission announced that it is going to spend 10 trillion Euros for EU defense. It intends to spend the money within the EU to build military equipment and weapons instead of buying from the US to boost the European military-industrial complex. In June 2025, Donald Trump increased tariffs up to 50% on EU steel and aluminum imports.

Joe Biden was involved in destroying the Nord Stream cheap gas supply pipeline, causing an increase in European cost of electricity supply. German industries are badly affected. Europeans need to import higher prices of gas and oil from the US. Donald Trump also wants to take over the sovereignty of Greenland against the wishes of the EU. The current relationship between the US and EU is badly fractured and financially the EU and USA are not in good shape. The trade war and tariff war instituted by Donald Trump prompted the Fed in March 2025 to announce heightened inflation, unemployment and slow growth. This would retard the global economy.

Many American politicians play money politics. They depend on the financial support of big businesses and the oligarchs during the election campaign. Many politicians want to remain in the Senate even when they reach 80 years old. The annual US government budget deficit approval depends on the Fed printing more fiat money out of thin air. The US reaps immense financial benefits from other countries when the Fed raises the interest rate. The US regularly exports hot money with very low interest rates to other countries and later raises the interest rate.

Many industries in the US have migrated abroad for many years to make bigger profits, and domestic industries are gradually disappearing. US manufacturing contributes only 10.3% to the US GDP. An important source of the economy now comes from the virtual economy , weapon manufacturing, petrodollars and international currency exchange fees. In 2020 , the virtual economy contributed nearly $2.6 trillion in values to the overall US economy. Many industrial products made in the US depend on the supply of components from China. Without China's supply chain support, the US manufacturing sector will be hampered.

The US government wants all US companies to move back from China, but it will not be able to do so unless the US makes significant investments in supply chain infrastructure. Because the cost of production in the US is too high, domestically produced goods are not competitive with foreign imports. Every product requires global supply chain cooperation to have any hope of reducing manufacturing costs. Trump's unilateral trade policies have broken global supply chains. BRICS members begin to trade using their own currencies, avoiding the traditional US dollar This would lead to the American financial banks losing a large amount of foreign exchange fees.

The US is a conflict-ridden society. Politically, the U.S. is alienating other countries through its long-arm jurisdiction, unlawful sanctions, imposition of tariffs, creation of animosity between neighboring countries, and the CIA's dissemination of misinformation and propaganda regarding other nations' leaders and governments. This is not helpful to promote the US as a friendly and dependable country for a healthy diplomatic relationship.

The confrontation between the two US political parties, the widening income of the rich and poor, the extreme left and right public opinion, the gun violence, the racial conflict between the black and white people, the religious conflict, the influence of the media, the insecurity of the people at the bottom, the judicial inequality, the highest crime rate per capita in the world would produce internal uprising in the future. American politicians are experts in creating fake hopes to cheat the masses with "American dream, protector of human rights and personal freedom and equality", but in reality the politicians lie, cheat and steal as exposed by ex-CIA Director Mike Pompeo. The US homeless population is rising and so is the wide gap between the rich and the poor.

The US created global terrorism, which was unheard of prior to September 11, 2001. It was an isolated incident of attack on the World Trade Centre by disgruntled Muslims who took revenge on the USA who supported Israel with inhumane treatment of Palestinian people. The US used the isolated incident as a world-wide terrorism and urged every nation to fight terrorism with the help of the USA. In the process of joint military operation against terrorism, the US begins to sell weapons to the world.

George W. Bush declared invasion of Iraq in March 2003 as he announced to the world Iraq possessed weapons of mass destruction. But the US later found no evidence of weapons of mass destruction in Iraq. The US killed over 185,000 Iraqi civilians. 4500 American lives were lost, and some 2 million Iraqis were displaced from their homes. In 2025, Donald Trump created a fentanyl war with China, Canada, and Mexico by complaining that these three countries have created national security risks by exporting fentanyl (a drug approved by US FDA as a pain relief drug) into the US. Using this disinformation as a reason that he raises import tariffs on these countries. The US is a well-known country for creating and distributing disinformation.

The two US political party politicians' acrimonious relations accelerate disunity in the society. Corruption amongst politicians in the deep state is rampant and cannot be eradicated. In 2024, the US national debt will exceed $35 trillion, and this figure is rising with inflation in tow. The US government may decide not to pay back its creditors the principal, but only to pay the interest, which has already reached $1 trillion. How long this US financial debacle will last is an open question. The global de-dollarization process is accelerating, deeply affecting the U.S. financial market; American power is destined to wane with its incessant warmongering, disruptive global trade and supply chain policies, and ever-rising national debt and inflation. The income gap between the rich and poor continues to widen, with the homeless population increasing.

Chapter 2: Political System in China
2.1 Past Imperial Political System

China is a huge country, covering an area of 9.6 million square kilometers. It is the world's second-most populous country, bordering 14 countries. China serves as a home to 56 ethnic groups. The Han group is the largest, making up over 92% of the population. People in each region of China from the North to South speak their own dialect, although all of them use the standard common written language. They all have different customs, traditions, and beliefs. Hence, it is very difficult for the central government to administer and manage such a diverse population of different people. The official communication with all 56 ethnic groups is very difficult, a situation often not appreciated by people in other countries, who tend to think China has a homogenous population and culture. They all think China has a homogenous population and culture.

Qin Shi Huang (259 – 210 BC) was the founder of the Qin Dynasty and the first emperor of China. He was instrumental in unifying the people across a vast territory that was home to many people with different languages. During his reign, he accomplished the most important achievement, which was the standardization of non-alphabetic written script across all of China, replacing the previous regional scripts. This script was simplified to allow faster writing, useful for national communication. The new script enabled parts of the empire that did not speak the same language to communicate easily. The Qin Dynasty also standardized weights and measures, with bronze standard models for measurements that were used nationwide in daily commercial activities. In conjunction with this, bronze coins were created to standardize money across the regions. The Qin Dynasty was able for the first time in Chinese history to unify the various warring states in China.

During the Han, Tang, and Song dynasties, China had the largest GDP in the world, proving that the Chinese had the skill, technology, economy, and security systems to rule the vast country many thousand years ago. However, every dynasty came to an end and was replaced by another dynasty. The collapse of every dynasty was mainly due to imperial official corruption and incompetent emperors. Chinese history proves that a country's prosperity or regression depends on the quality of the emperor and his administrative officials, and not political doctrine.

The Chinese people traditionally relied on the emperor for safety and social stability. The emperor had an annual budget for maintaining the country's administration. Hence, the imperial court issued a decree to collect tax to pay for the cost of administration, social infrastructure maintenance and military expenditure. The Han, Tang, and Song dynasties had competent emperors and efficient imperial court administration to achieve high economic growth and agricultural production. The last imperial government in China ended by a revolution in 1911 when the corrupt Qing Dynasty was toppled by the Kuomintang Party . However, the party was unable to rid the warlords and oppressive landlords, leaving the farmers to continue to live in primitive conditions. Later, the poor farmers helped the Communist Party topple the corrupt Kuomintang government, which later retreated to Taiwan in 1945.

China has a long history of five thousand years with various dynasties. Each successive dynasty brings improvements and progress in politics, economy, culture, invention, innovation, and other aspects of social development. According to records, there were a total of 83 dynasties and a total of 559 emperors. The Imperial court rules the country by using the common Chinese language, although different regions speak different dialects. In Chinese history, unlike European countries, China never colonized any country. In fact, all the emperors built great walls to prevent foreign invasion.

2.2 INVENTIONS IN CHINESE 5000-YEAR HISTORY

In the long history of Imperial rule in China, there were many inventions produced by the public without getting the benefit of being the owner of intellectual property. Here is a long list of inventions during various dynasties:

- Acupuncture: from 3rd to 2nd centuries BC.
- Armillary sphere, hydraulic-powered: Han Dynasty (202 BC – 220 AD)
- Banknote, Paper currency: Tang Dynasty (618–907)
- Bellows, hydraulic-powered: Han Dynasty (202 BC – 220 AD)
- Belt drive: 15 BC
- Blast furnace: Han Dynasty (202 BC – 220 AD)
- Bomb: Jin Dynasty (1115–1234)
- Borehole drilling: Han Dynasty (202 BC – 220 AD)
- Breeching strap: Warring States (481–221 BC)
- Brine mining: invented around 500 BC
- Bristle toothbrush: about, 1498
- Bulkhead partition: Song Dynasty (960–1279)
- Candle clock: 6th century AD.
- Cannon: 13th century.
- Cast iron: Zhou Dynasty (1122–256 BC)
- Chain drive, endless power-transmitting: Song Dynasty (960–1279)
- Chinese soup spoon: Shang dynasty in 2nd millennium B.C.
- Chopsticks: Shang Dynasty, 11th century BC

- Chromium: about 210 BC.
- Civil service: Sui Dynasty (581–618)
- Co-fusion steel process: 6th century AD.
- Coke as fuel: The Song Dynasty (960–1279)
- Color printing: Yuan Dynasty. About 1000 AD.
- Contour canal: First Emperor Qin
- Counting rods: the 2nd century BC
- Crossbow: 2000 BC
- Metal Casting : Han Dynasty (202 BC – 220 AD)
- Dental amalgam: Tang Dynasty (618–907 A.D.)
- Drilling rig: Han Dynasty, (202 BC – 220 AD)
- Field mill: Late Zhao (319–351)
- Firecracker: 200 BC
- Fishing reel: Southern Song (1127–1279)
- Folding screen: Eastern Zhou Dynasty (771–256 BC)
- Forensic entomology: Song Dynasty (960–1279)
- Fuses: 10th- 12th centuries
- Gas cylinder: Tang dynasty (618–907 A.D.)
- Gas lighting: around 500 B.C
- Gear: 4th century BC
- Heavy moldboard iron plow: 500 BC
- Helicopter rotor: 400 BC
- Horse collar: Northern Wei (386–534)
- Horse harness: State of Chu, 4th century BC
- Hygrometer: Western Han dynasty
- Inoculation, treatment of smallpox Song Dynasty (960–1279),
- Match: Northern Qi (550–577)
- Multiple-tube seed drill: Han Dynasty (202 BC – 220 AD).
- Oil refining: 512 AD- 518 AD
- Oil well: 347 AD

- Paper packaging: second century BC.
- Pendulum: Han Dynasty (202 BC–220 AD)
- Pig iron: Zhou Dynasty, 5th century
- Pontoon bridge: 9th or 8th century BC
- Porcelain: Tang Dynasty (618–907)
- Rockets: Song dynasty (960–1279)
- Rocket bombs, explosive payloads: Ming Dynasty (1368–1644)
- Rocket launcher: 11th-century
- Rotary fan, manual and water-powered: Han Dynasty (202 BC–220 AD)
- Salt well: around 2250 years ago
- Seismometer, Han Dynasty (202 BC–220 AD)
- Sky lantern: 3rd century BC
- Snow gauge: Southern Song dynasty (1127-1279)
- Solid-propellant rocket: Song dynasty (960–1279)
- Soybean oil: 2000 BC
- Soy sauce: Western Han dynasty (206BC-9 AD)
- Suspension bridge using iron chains: Yongle Emperor (1402–1424);
- Tea: 2nd millennium BC
- Thyroid hormones to treat goiters: 1st century BC
- Tofu: 179–122 BC
- Toilet paper: in the year 589
- Tung oil: Song Dynasty (960–1279)
- Well drilling: 347 AD
- Wheat gluten: 6th century China
- Wheelbarrow: 118 CE
- Winnowing machine, Rotary fan: Han dynasty (202 BC–220 AD)
- Wrought iron: Han Dynasty (202 BC–220 AD)

- Silk: 6000 years ago
- Umbrella: around 3500 BC

In 806 AD, China invented paper money to replace silver coins, as silver coins were too heavy to carry when dealing with large business transactions. Banknotes were issued in Chengdu in the early years of the Northern Song Dynasty (998 to 1003 AD). The "Jiaozi" was the earliest paper money in the world. Paper bills were originally used as privately issued bills of credit or exchange notes. A merchant could deposit his cash in the firm, receiving a paper "exchange certificate" which he could exchange for silver coins in other places of the same firm. It was a great achievement in the history of mankind in using paper money for trade and payment. Now, digital money is used by Chinese people to pay for purchases of goods and services via Alipay or WeChat Pay to settle the payment. Payment is done by using a mobile phone. Soon, paper money will become obsolete. Use of digital money saves the banks massive amounts of costs in labor, administration, material and time. The people in the US use credit cards for purchase of goods and services and can easily over spend beyond their means and sustain debt. Chinese people using digital payment can avoid over spending.

In 2025, China has invented and developed a prototype of ultra-longevity carbon-14 (C-14) nuclear battery, named Zhulong-1. It can convert radioactive decay energy into electricity. It has a designed lifespan of 50 years for use in pacemakers and support networks of trillions of sensors for the Internet of Things and interstellar spacecraft and weaponry.

2.3 POLITICAL SYSTEM CHANGE FROM IMPERIALISM TO DEMOCRACY

Confucius (551 – 479 BC) has been a revered scholar and philosopher in China. His philosophical teachings emphasize personal and governmental morality, harmonious social relationships, righteousness, kindness, sincerity, and a ruler's responsibilities to lead by virtue. His teaching emphasizes that the strength of a government ultimately is based on the support of the people and virtuous conduct of the ruler. Throughout Chinese history, his philosophy had become the imperial rulers' guiding principle of governance. Successive dynasty rulers followed his teaching with success in the state administration. Corruption by court officials and incompetent emperors led to the demise of the last Qing dynasty, which was overthrown by the revolution led by Sun Yat-sen of the Kuomintang Party (KMT) in 1911, ending 267 years of imperial rule. He established the first democratic republic in China, thus ending more than 2,000 years of absolute monarchy in China.

Sun was elected as the Provisional President of the Republic of China on 1 January 1912. Due to intense political infighting, his presidency was handed over to Yuan Shikai who was the leader of the Beiyang Army. He tried to replace the Republic with his own imperial dynasty. When he died in 1916, the country was plunged into disarray , creating many warlords around the country. Later, Sun had tried to get Japan, Britain, France, the US and other countries to help China's reconstruction in the early days but without success. On March 12, 1925, Sun died, and the party leadership was handed to Chiang Kai-shek. During his administration, Chiang had a tough time dealing with three major events, namely eliminating all the warlords in China, fighting Japanese invasion and civil war with the Communist Party (CPC).

Japan invaded the Chinese province of Manchuria in 1931. By 1937 Japan controlled large parts of China, and war crimes against the Chinese became commonplace. It was a difficult period for Chiang to fight on so many fronts with limited resources. The Second Sino-Japanese War was the war fought from 1937 to 1945. Chiang and Mao agreed to stop the civil war and unite to fight the Japanese invaders. During WW II, China received substantial military aid from the US , Russia, and UK. On 2 September 1945, Japan formally surrendered following the atomic bombings of Hiroshima and Nagasaki by the US. The war resulted in the deaths of around 20 million Chinese people, mostly civilians, including more than 300,000 civilians murdered by the Japanese soldiers in Nanjing.

After the Japanese surrender in 1945, both KMT and CPC continued the civil war. With massive support from the rural farmers, whose living conditions not only did not improve under KMT rule but also constantly being harassed by warlords and oppressed by landlords, the CPC finally defeated the KMT Government. Chiang retreated to Taiwan, where his KMT government continued to function. Chiang's authoritarian rule, hyperinflation, lack of support from the huge rural population and corruption in the government led to his government's downfall. On 1 October 1949 Mao Zedong established the People's Republic of China (PRC). He had promised the farmers that he would give them a free plot of land to farm for their help in defeating the KMT government.

Mao Zedong's government faced tremendous challenges to rebuild the country devastated by long periods of wars. The treasury was empty without foreign exchanges that had been taken lock, stock, and barrel by Chiang, and many professionals migrated overseas. The civil service was in disarray. The food supply was in a dire situation. Mao Zedong inherited a bankrupt country, trampled by foreign invaders and corrupt politicians. Millions of people lived in a hellish condition.

2.4 Chinese Political System Evolution

The CPC government endeavored to serve the people by raising the living standard and eliminating inequality in the society. Mao Zedong emphasized that men and women are equal. His government was faced with lack of foodstuff and immense infrastructure destruction due to many years of wars. Many remnant thugs and warlords are still harassing the poor farmers in the rural areas. To manage effectively a huge country with 56 ethnic groups, a population of 541 million people with thousands of years of feudal tradition was no easy task for any government. The GDP of China was similar to the poor country in Africa. Mao Zedong applied tough measures to unite the whole country.

Mao established social justice to all citizens, irrespective of social background. Simultaneously, he ensured there was sufficient food and housing for everyone. His draconian measures raised many objections from some comrades in the Communist Party. Due to the US anti-Communism foreign policy and US veto power in the UN, the PRC government was not recognized by the UN, and hence received no aid from the UN. The only foreign country that offered aid was the Soviet Union. Mao did try to implement many novel ideas to improve agriculture production but failed for varied reasons, one of which was due to lack of technical know-how.

Cooling relations with the US began in 1972 when President Nixon visited China and later established diplomatic relations with China. Deng Xiaoping in 1978 began to open up China to the outside world and introduced a market economy. With the help of overseas Chinese diaspora establishing factories in China, manufacturing consumer products exporting to other countries, the Chinese economy began to progress. By 2012, China achieved the second-highest GDP in the world, surpassing Japan. The Chinese government had lifted 800 million people out of poverty. Such achievement is unprecedented in the history of mankind. The government relaxed travel restrictions, allowing Chinese citizens to travel freely to foreign countries. Chinese outbound tourists become the highest in the world. According to a report, 155 million Chinese spent US$133.8 billion on outbound trips in 2019. China attained the title of "Factory of the world" because of its global trade with more than 120 countries. China has adopted capitalism amidst its socialism. True Communism disappeared from China.

Since Donald Trump became the US President in 2017, geopolitics has become more intense due to his hardline foreign policy, provoking conflicts around the world and starting a crusade of high tariffs on every country around the world. The wars in the Middle East and the 2008 US financial crises had made the world very unstable. However, China was able to weather the storm. The US has been pressing China to denounce Russian invasion of Ukraine, but the response from China has been asking both Ukraine and Russia to negotiate for a cease fire.

Due to the invasion, Russia has been sanctioned by the US and EU, with many US and EU companies exiting the Russian market. As a result, the trade between China and Russia increases many folds. The Chinese RMB is used in the trade between China and Russia. China has been steadfast not to get involved in the wars created by the US in the Middle East countries. On the diplomatic front, China was able to convince both Saudi Arabia and Iran to establish diplomatic relations, despite both countries being longtime rivals for years in the Middle East.

China's political influences have expanded rapidly by the Belt and Road Initiative, BRICS, SCO, AIIB and RCEP. The Chinese government has done a good job in lifting 800 million people out of poverty in the past 70 years. It provided modern transport systems of high speed rails, highways, and airports around the country. Previously, the annual arduous long home road trips from the cities made by migrant workers on their motorbikes during Chinese New Year period suffering from wintry cold and rain have now become history. The Chinese government even helps the migrant workers taking the high speed rail home with affordable fare.

Many farmer's living standards have improved by the government program of building good roads to every village so that the farmers can either sell their produce through online marketing on social media like Douyin, WeChat and Weibo or establish homestay business. The income of many rural residents has increased so much that they can afford to build modern houses on their own farm land. China's international political standing is on par with the US. The US realized that smearing the reputation of China in the social media has no effect on China's peaceful image.

2.5 US MILITARY THREAT AGAINST CHINA

Currently, the US has 750 military bases stationed in 85 countries and regions. What is the purpose of the US stationing so many military bases in so many countries? The US government said that bases are meant for maintaining peace for other countries. On the contrary, these bases have been used by the US to invade other countries like Iraq and Afghanistan, Yemen and Syria. It costs a mammoth amount of resources of manpower and finance to maintain and support the bases' expenditure. The US defense budget reaches US$893 billion in 2025. Every US ally in Europe, Middle East and Far East has to contribute financially towards supporting the US military base expenditure. Stationing troops in overseas countries saves a lot of money for the US government compared with stationing them at home. In addition, every ally needs to purchase military weapons from the US. The price of the weapons is determined by the weapon manufacturers. The US military operations in the Middle East had gained a large amount of oil benefits. Both Japan and South Korea have to contribute towards the operational expenditure of US military bases in their countries.

Since 1949, the US has been deploying its aircraft carrier and other naval vessels patrolling the South China Sea on the pretext of maintaining freedom of navigation in international waters. However, the actual purpose is to create tension in the Taiwan Straits to support Taiwan politicians who call for independence. The US seizes the opportunity to sell weapons to Taiwan on the pretext that the US would protect Taiwan in case of war with mainland China. Once the weapons are sold, the US will send military personnel to train the Taiwanese personnel and also station so-called military advisors to advise the Taiwan government on foreign affairs dealing with mainland China.

The US also invited other NATO members and Australia to participate in US naval exercises in the South China Sea regularly, posing a military threat against China. It also set up a military alliance with the UK and Australia to confront China by setting up AUKUS alliance to 'protect' the shipping lanes in South China Sea which is mainly used by China. In 2022, the US convinced Australia to cancel the diesel powered submarine contract with France and replace it with a contract to buy 8 US nuclear submarines at a cost of $ 268-386 billion. The Australian nuclear submarines are to help the US in case of a war with China in the South China Sea. The long range nuclear submarine is for offensive military operation. The original French contract diesel submarines are adequate for defense purposes, and they are much cheaper and easier to maintain.

The US has been openly helping the Hong Kong young activists financially to stage bloody and violent demonstrations against the Hong Kong government in 2019-2020. During the demonstrations that extended for many weeks, the young demonstrators waved American flags, damaged the legislature building, public buildings and occupied the international airport, causing safety concerns to arriving and departing foreign passengers . The demonstrators killed one pedestrian. The activists occupied and used a university laboratory to produce firebombs. Joshua Wong Chi-fung and Jimmy Lai were the leaders of the civil riots, and they had the audacity to go to the US Congress to appeal for assistance to crush the Hong Kong government. Most activists were teenagers, and they even forced their parents to support their public violent demonstrations. The Chinese government exercised extreme restraint, not to send the PLA soldiers stationed at Hong Kong to quell the riots.

The intense demonstration stopped as soon as China introduced a national security law to prohibit violent demonstration and replaced all the foreign court judges who sympathized with the demonstrators. Many demonstrators were caught and charged in court but were released immediately by the pro-demonstrator judges. It was an open secret that the CIA supplied financial aids and organized the demonstrations that were well planned and organized remotely in the US. The long demonstration caused many foreigners to flee Hong Kong, as the situation was getting very tense and they felt unsafe. The civil unrest would have succeeded in toppling Hong Kong had not China passed a national law that prohibits violence demonstration. CIA agents' job function is to sponsor violent demonstrations in a foreign country by creating color revolutions in several countries. The CIA's famous slogan for waging violence and revolutions is "defending democracy" for every country in the world. CIA covert actions are not answerable to the US Congress. In a sense, the CIA is a criminal organization, not unsimilar to a terrorist organization.

2.6 CIA Anti-China Propaganda

The CIA , the world's largest propaganda organization, world renown troublemaker, hatred mongering organization, has an anti-China unit that receives US$1.6 billion budget in 2024 for spreading propaganda and disinformation about China. This department sends out anti-China messages to countries friendly to the US, for them to spread worldwide. The CIA pays anyone who is a "fake news creator" from around the world to work for the CIA to smear China. It is cheaper for the US to create chaos and destroy China with money than with costly military action.

In addition, the other four US allies of English-speaking countries, Australia, Canada, the United Kingdom and New Zealand, can also be persuaded to join the CIA's anti-China campaign and impose trade sanctions against China. Australian Strategic Policy Institute (ASPI) advocates for Australia to become a partner of US foreign policy to constrain China. It alleged China committed genocide of Uyghurs in Xinjiang. It publicized the conspiracy theory that the Chinese government was responsible for the COVID-19 outbreak in Wuhan. ASPI is being used as a conduit for hardline US policy towards China.

The current anti-China theme focuses on three aspects of defamation against China: a) the genocide of the Uyghur people in Xinjiang. b) The National Security Law restricting democracy in Hong Kong. c) China's threat to Taiwan's democracy. The ultimate aim of anti-China is to cause civil unrest in China until China disintegrates. Former CIA Director and former Secretary of State Mike Pompeo once said, "At the CIA, we lie, we cheat, we steal." Another anti-China approach came when the CIA commissioned two agencies to produce disinformation about Xinjiang. Adrian Zenz is a German who is a religious leader with extreme religious views. The BBC, the Guardian, the New York Times, the Washington Post and the Economist, among other US and British media outlets, all commissioned Zenz to write a sinister report on Xinjiang. He ended up writing an article about the abuse of Uyghurs in Xinjiang by the Chinese government. His articles were published in all the British newspapers. After the UK left the EU, the UK had become an active, fervent supporter of the US anti-China policy.

Other false messages were created by another man named Omar Kanat, the Chairman of the World's Uyghur Congress (WUC), who is stationed in the U.S. The WUC receives significant funding from the National Endowment for Democracy (NED) in the US. NED is another branch of the CIA. NED's job is to create insurgencies and chaos in countries that do not fully submit to the will of the US. The false news created by the CIA agents are reported by the US media every day, and repeated in the West media, thus affecting China's reputation globally, causing people in many countries to dislike China.

The US propaganda war against China has been a perpetual political campaign since 1949 when PRC was established by Mao Zedong. The US will never want to see the day when China climbs over the US to become the strongest economy in the world. In a way, it is similar to the US propaganda war against Russia as a perpetual enemy. This explains the reason that the US has 750 military bases in Europe, Middle East and Asia. The military bases are also used as a propaganda tool to help the US sell weapons to other countries that their security are "threatened" by the evil regime in Russia, China, and Iran proclaimed by the CIA. In addition to the propaganda war against China, the US now has a trade war and tariff war against China as well as against its allies. It is a strange human behavior of the US politicians who think that smearing and containing China will eventually cause the downfall of the PRC, which has the full support of the whole population and possesses no less powerful modern weapons than the US weapons. It will be a miracle that the US politicians can annihilate China from the earth when it cannot even defeat a poor country like Afghanistan in 20 years.

2.7 POLITICAL RIVALRY BETWEEN CHINA AND USA

The US government relentlessly demonizes, contains and sanctions China, regarded by the US as an enemy. President Obama intended to bring many Asia-Pacific countries together to contain China through the TPP (Trans-Pacific Partnership). At present, China is the biggest economic rival of the US, which is bent on preventing China's rise. The US realizes that attacking China with aircraft carriers in the South China Sea will not work because China can deploy swarms of missiles and drones from the shore, and the slow-moving navy ships become sitting ducks. In addition, the Chinese DF-27 missile, equipped with a hypersonic glide vehicle, is described as the world's first anti-ship ballistic missile (ASBM) or "carrier killer". It is capable of deploying a 600 kg payload with a minimum range of 500 km and a maximum range of 8,000 km. The DF-27 warhead is maneuverable, with a 3-meter point accuracy.

If China were to destroy the US supply warships first, the aircraft carrier would have no ammunition replenishments to continue fighting a war of attrition. If and when China sinks a US aircraft carrier with 7000 sailors and 100 jet fighters onboard, the US would immediately declare a truce to stop the war. The US war strategy has not changed since WWII, using aircraft carriers as the main weapon to fight a country without comparable firepower. Its aircraft carriers in the European region to frighten Russia for decades have no relevance since Russia possesses nuclear and massive defense weapons. Aircraft carriers are good for war against small and weak countries with little weapons to defend their security.

In January 2025, Trump's choice of Pete Hegseth as Secretary of Defense reveals Trump's utter failure of employing a competent cabinet candidate when Pete cannot utter one country that belongs to ASEAN. Instead, he picked Japan and South Korea as ASEAN members during the Congressional hearing on his appointment. In March 2025, he texted about US military plans to kill a Houthis militant leader in Yemen to Signal chat that included Atlantic Editor-in-chief Jeffrey Goldberg. The exposed military plan is meant to be classified. As a Defense Secretary on the job for less than three months, his slovenly action shows his mediocre professionalism. Employing an amateurish person in an important government cabinet position is a sign of weak governance. How can he expect to win a war with China or any other country? Similarly, Vice President JD Vance said the US borrowed money from Chinese peasants to buy things Chinese peasants manufactured. He probably refused to accept that the Pakistan Air Force, on May 7, 2025, using J-10C fighter jets made by Chinese manufacturers, shot down three Indian Rafale fighter jets made in France. The J-10C cost $40 million, compared with $240 million for the Rafale jet.

US Defense Secretary Pete Hegseth and JD Vance probably did not know that the Pakistani pilots could shoot down the Rafale because of the latest air battle digital radar network tactical air battle ecosystem designed by the Chinese advanced system-center, instead of the traditional platform-center in a dogfight. The new Chinese tactical closed-loop ecosystem for target discovery to missile hit uses ground-based long-range radar integrated with an AWAC (airborne warning and control system) radar. This system determines an enemy fighter jet's position from a long range, then relays its exact position to the J-10C radar, which finally fires the PL-15E supersonic missile at the radar-locked enemy jet. The AWAC radar guides the missile precisely to complete the end-to-end guidance through the BeiDou satellite navigation system, forming a closed-loop highly integrated digital tactical system. The fighter jet that sees and shoots first before the enemy will score first. The Rafale jets were shot down in Indian territory. The firing of missiles that are beyond visual range. The Rafale fighter pilot relies on his stand-alone radar performance that has a shorter detection range. The Indian pilot does not have such an advanced digital radar ecosystem.

China has been using IT and AI to innovate a new digital aerial battle ecosystem. The latest Pakistan-India air battle involving more than 100 fighter jets proves the latest Chinese dogfight ecosystem of shifting from "weapon generation competition" to "system effectiveness competition". Many American and Western military officials are unanimous in belittling the J-10C fighter jets, having no battlefield experience to prove its worth. In general, American and Western military personnel look down on Chinese military equipment and in general their thinking is still based on traditional military hardware and traditional military battle ecosystem. The US proxy Ukraine war is still fought with old weapons and traditional hardware, although both sides have started using drones as new weapons. Can the US win a hot war with China in the South China Sea region using traditional weapons and hardware against China?

In another war scenario, if a war between China and the US begins, the Chinese DF-27 ballistic missile can directly strike targets such as the US territory of Guam. It can also use the BeiDou navigation aiming system to hit the propeller of a US aircraft carrier, which would disable it. In naval warfare in the South China Sea, the US is no match with China's land-based military power. American weapon spares support is too far away from the South China Sea, and hence the US cannot fight a war of attrition with China. The US is also concerned that Russia and North Korea would help China attack the American naval fleet. No country in Asia will dare to join the US to attack China, as Asian countries have suffered inhumane miseries caused by Western and US colonization.

The US warships patrolling the South China Sea are more of a show without significant military value in actual warfare. In 2024, China tested an ICBM that could reach the US mainland. The US is also mindful of losing a hot war with China that will generate earth-shattering news around the world that the US is a paper tiger. Continuing US military bases in Japan and South Korea will have no military significance. China will instantly become the superpower in Asia. Perhaps, when the situation is ripe, China may provoke the US into a war in the South China Sea to end the prolonged US containment strategy, which has yielded no realizable benefit to US national security.

China is about to start building an Asian alliance, boosting China's political power to a new height that would shock the US. According to the United Nations 2023 statistics, Asia has 48 countries, a population of 4.7 billion (60% of the world), and a GDP of $41.36 trillion. So the Asian Alliance is a powerful international force. The establishments of BRICS, SCO and RCEP have already enhanced China's power in the international political arena, whilst the US political influence in Asia is waning. China operates or has ownership of 115 ports around the world. China's GDP growth rate reached 5.2% in 2023, much higher than the EU's 1.2% and the US's 2.4%. The interest payment on US Treasury bonds has reached $1 trillion, and many foreign countries have begun to dispose of their US treasury bonds to avoid the US defaulting on repayment on the principal. The US high inflation, national debt and unemployment do not bore well for the US economy, coupled with the continuous financial support in terms of tens of billions to Ukraine and Israel that would drain the US treasury dry.

The process of de-dollarization around the world is underway, causing the US dollar demand and circulation to reduce dramatically in the world. The US political dominance begins to wane in many countries as it seeks to create war actively compared with China building friendship with other countries especially those in BRI. Many countries see through the double standard of American human rights when the US has a detention camp in Guantánamo Bay in Cuba where injustice, abuse and disregard for rule of law against the prisoners occurred.

The continuous military and financial support for Israel—enabling it to commit heinous crimes against humanity, including the killing of innocent Palestinian women and babies and the bombing of U.S.-sponsored humanitarian aid supplies and UN workers—reveals the hypocrisy of American human rights and freedom policies. Donald Trump starting a worldwide crusade of tariffs in 2025 against every country including its allies like Canada, EU, Australia, Japan, and South Korea has created a tsunami of strong protest against the US. Consequently, more countries want to join the BRICS. Suddenly, China has become a favorite country with which to do trade. Suddenly, the CIA propaganda stops describing China as Communist China, which has a subtle meaning of being Evil China. The CIA cannot expect US allies like Canada, Australia, and the EU to trade with 'Evil China,' but trading by the US with 'Evil China' is acceptable. The US needs high-quality and inexpensive goods made by Chinese manufacturers, as stated by VP JD Vance.

2.8 CHINA'S DIPLOMATIC RELATIONS WITH MIDDLE EAST COUNTRIES

The "25-Year Cooperation Plan for the Iran-China Comprehensive Strategic Partnership", signed by China and Iran in Tehran on March 27, 2021, is a 25-year cooperation agreement. It aims at further developing Iran-China relations, with China investing US$400 billion in the Iranian economy. In March 2023, China brokered a surprise détente between Iran and Saudi Arabia. This follows years of intense rivalry between the two countries that had destabilized several Middle Eastern countries, including Iraq, Syria, Lebanon, Yemen, and Bahrain. Officials in Tehran and Riyadh had cited their countries' deteriorating relations with the US as one of the main reasons for their policy shift. Saudi Arabia and Iran had restored diplomatic relations in a deal brokered by China. The agreement marked a diplomatic victory for China in a region long dominated by US geopolitics. Such geopolitical development together with the expired agreement on petrodollar between Saudi Arabia and USA indicates the waning power of the US in the Middle East and the corresponding rise of China political influence in the region.

In 2023, trade between Saudi Arabia and China reached US$106 billion. Saudi Arabia also began to accept Chinese yuan currency to buy oil, moving away from the petrodollar. In 2023, Saudi Arabia, Iran, Ethiopia, Egypt and the United Arab Emirates applied to join the BRICS and became BRICS members in January 2024. Members begin trading in their own currencies instead of the US dollar. China is a leading country in BRICS. Many of the organization's members are large oil producers, using their national currencies to trade. This would seriously hurt the US financial market and directly reduce the

US dollar foreign exchange earnings. China's political influences are on the rise in the Middle East and will strengthen trade, social and economic relations between China and the Middle East. At the same time, America's political influence in the Middle East is declining.

China has signed Belt and Road (BRI) cooperation agreements with 22 Arab nations. This framework has yielded over 200 large-scale projects. The 1st China-Arab States Summit took place in 2022. Saudi Arabia, the United Arab Emirates, Egypt, and Tunisia have announced that the Chinese language will be included in their national education systems. There are more than 20 Confucius Institutes in Arab countries and hundreds of schools offering Chinese courses. China builds the first high speed rail in Saudi Arabia. Many government leaders from the Middle East countries had visited China in the past few years to discuss various subjects of mutual interests. China is a country friendly with Muslims whereas the US has invaded many Muslim countries like Iraq, Afghanistan, Yemen, Syria, and Libya.

2.9 DIPLOMATIC RELATIONS BETWEEN CHINA AND AFRICAN COUNTRIES

At various international conferences held in Africa, many African leaders reiterated that whenever the leaders of the US and the European Union visited African countries, they talked about only three topics, namely China, debt trap and military alliances. When Chinese leaders visited Africa, the talk was about infrastructure and investment. African leaders reminded European leaders about Africa's past history of being colonized by them. During the period of European colonization, Africa's natural resources were plundered by colonial powers without any compensation to the local people. At present, only China has built dams, power plants, seaports, airports, railways, roads, schools, hospitals, energy and water supply, and sanitation facilities across African countries through the Belt and Road Initiative, benefiting 49 countries. Investments over the past decade have yielded tangible results. Various projects have been completed to help African countries address key barriers to accessing finance for economic growth and social infrastructure improvements.

Many African leaders have visited China in the past decade. They all have ridden on the Chinese high speed train with maximum speed of 350 km/h and appreciate the smooth ride in seats resembling aircraft seats with lots of legroom. African leaders are impressed by China's technical advances, and they would like to take a leaf from China's experience and policy. China offers scholarships to African students to study in China. With China's financial and technical assistance, Africa's economy has improved significantly. China's diplomatic relations with

African countries have solidified over the years. China's sphere of political influence in Africa has surpassed that of European countries and the US. African countries remember the atrocities committed in Africa from the days of slavery trade by the British Empire and plundering of African natural resources by European countries like France, Germany, The Netherlands , Spain, Belgium , Italy , and Portugal. African countries possess a lot of natural resources, and with technical and financial help from China to develop them, China can assist their economic development.

Chapter 3: Chinese Economy
3.1 Reconstruction of Chinese Economy Post PRC Establishment

It *was* a herculean task to feed and provide shelter to a population of 542 million people in 1949 with the country devastated due to decades of fighting and a large destitute rural population. Mao Zedong's initial tasks were to eliminate all the warlords and landlords that had oppressed the poor farmers for decades. Secondly, he had to unite the whole country by removing activists who opposed the Communist Party and to implement measures to provide equality for the people and eliminate century old superstition. He needed his long-time comrades to help him achieve his goals. Differing opinions abounds as to how to achieve the final goals. The Japanese invasion had destroyed the country's infrastructures, and the local food production was inadequate to meet requirements.

After WW II, every country, except the US, suffered immense economic and financial difficulties. PRC could not get the UN to help because PRC was not recognized by the UN. Neither could the PRC get help from the IMF. The initial period of PRC implementation of a civil service organization to cover the whole country was arduous. The problem was compounded by the low education level of many administrators, especially in the poor regions and towns. Many rural populations were illiterate. Effective official communication was hampered by different dialects spoken by various ethnic groups throughout the country. China in 1949 was a basket case. One important measure taken to ensure that there was no undue pressure of unemployment in the cities was to create a residence permit that controlled migration of farmers to the cities. Too many farmers migrating to the cities could create problems of housing, medical care, and jobs. Farmers migrating to the cities would reduce production of food crops.

Mao Zedong promised all the farmers to be given free land for agricultural purposes so that their livelihood could be ensured. A big proportion of China's population was the rural inhabitants. Production of food crops was inadequate due to lack of fertilizer, insecticides, and technical knowhow. Imports of foodstuffs like rice, wheat, and soybean were needed. Also, farming was frequently affected by bad weather like drought and floods. Overseas Chinese diaspora who sent foreign money to their relatives in China were used by the government to import essential supplies. Mao died on 9 September 1976 . He was credited with transforming China from a semi-colony to a world power by improving literacy, women's rights, basic healthcare and primary education. Up till now, Mao is revered as a national hero. Without him, China would still remain as a Third World country.

In 1978, Deng Xiaoping boldly carried out reform of the Chinese economy by applying market economy policy, opening up China to the outside world. He declared : "black or white cats are good cats if they can catch mice" and "making money is glorious". As China had no prior experience in capitalism , he applied the concept of "crossing the river by feeling the stones," attempting capitalism to improve the economy to meet the needs of ordinary people and improve their living standard. He convinced the party members that, together with reform and opening up, only can bring China's potential to fruition. He said without reform, the country would be doomed. As he wanted to learn how he could open up the China market to the world, he visited

Singapore to see how Singapore could achieve high economic progress without any natural resources.

Singapore Prime Minister Lee Kuan Yew encouraged him by saying if Singapore could achieve economic success with poor migrants from China, China could achieve better results as China had many more talented people than Singapore. He promised Deng that he would send the retired Minister of Finance to China to help formulate economic policy. He also helped China to facilitate foreign companies setting up factories in China by establishing Suzhou Industrial Park. For many years, Deng sent many civil officials to attend training courses in Singapore to learn Singapore economic development models. Many Chinese mayors and other officials had attended customized courses in a Singapore university. Such overseas training courses for Chinese officials had never happened before.

For his initial implementation of his market economy policy , Deng selected Shenzhen as the first trial city. Shenzhen was a city free from the encumbrance of party politics and close to Hong Kong, which is a free port and an international financial center for trade transaction settlement. Overseas investors could fly in and out of Hong Kong freely and easily. International fights between China and other countries were limited at that time. The initial investors mostly came from Hong Kong, Taiwan and overseas Chinese diaspora from South East Asia. They set up factories to export consumer products that required low skill. The initial exports included garments, shoes, children toys, household items, simple electronic products like calculators, and portable transistor radios.

The foreign investors had experience in dealing with international trade, rules, and regulations and standards. As expected, there were many teething problems initially faced by the foreign factory managements dealing with the semi-literate workers on quality issues and compliance with international rules and regulations. The local government officials were also inexperienced in dealing with labor and management disputes. Frequent friction had risen between the workers and management, as this was the first time Chinese workers reported to foreign managers with different culture and custom.

In May 1980 the Central Committee designated Shenzhen as the first (Special Economic Zone) SEZ in China. Shenzhen was a small fishing village with some 30,000 residents living barely above the subsistence level. Shenzhen's GDP has grown from 196 million yuan in 1979 to reach nearly 3.24 trillion yuan (US$477.4 billion) in 2022 which is larger than that of the Philippines of $404.4 billion with a population of 115.6 million compared with Shenzhen's population of 17.7 million. Shenzhen GDP per capita has grown 56 times within 43 years. Shenzhen has transformed from a fishing village to a modern, advanced and rich metropolis with thousands of high-tech company headquarters in operation. It is called China's Silicon Valley.

Shenzhen is a big city of immigrants, including many foreigners. Shenzhen's economic success is due to the migration of talented people from all over China to Shenzhen to open up domestic and foreign trade and enterprises. The Shenzhen government provides a free open market policy, welcoming investors from four corners of the world. The local government leaders have been very supportive to investors to set up factories and financial institutions to help build the city economy. How come the Philippines with a large population cannot achieve the same GDP as Shenzhen City when the Philippines has all the human, technical and financial resources?

Can any other city or country emulate the success of Shenzhen in a period of 43 years? The answer is not likely as many countries do not have long term economic planning as they change governments frequently. Each new government has its own short term agenda, but long planning for 30–40 years is not possible. Hence, Shenzhen economic achievement cannot be replicated in other countries as it has the best supply chain for electronic and consumer product manufacturing. Also, many politicians in other countries play money politics instead of concentrating on long term national development. Unless a miracle happens, Donald Trump's MAGA is a pipe dream within a 4-year presidency. It is just an election slogan.

3.2 Deng Xiaoping Economic Policy Achievement

China's rapid economic success is credited by many small and medium size private enterprises exporting to many countries around the world. Their products are competitive due to their prices being lower than competitors' products. In 2023 about 80% of Walmart's inventory in the US will be imported from China. China exports to the United States value was US$501.22 Billion during 2023, according to the United Nations report. China is the largest exporter of goods in the world, with a total export value in 2022 of US$3.71 trillion, which is higher than the GDP of the UK at $3.08 trillion. China became the factory of the world and reached the second-highest economy in the world in 2012 surpassing Japan.

The government has invested heavily in a myriad of industries. China is well known for green energy industries, including solar panels and wind turbines. Other globally well-known companies include Huawei for its communication equipment, Alibaba for its online shopping and Alipay digital payment, BYD for its electric vehicles and buses, CRRC for its high-speed rail construction, CATL for its batteries, DJI for its drones, Comac for its aircraft manufacturing, High-Flyer for its DeepSeek generative AI bot, JD for its largest online retailer in China, ByteDance for its TikTok, and Tencent Holdings Ltd for its video games and WeChat payment. All these industries require high-tech knowledge and application. China is able to send spaceships to orbit round the world and send space capsules to the far side of the moon to collect soil and rocks.

China is the leader in the global supply chain. Many manufacturing industries in the US rely on many raw materials supplied by China, and some of these raw materials cannot be sourced from other countries. The US navy ship builders import 40% of their semiconductor requirement from China. China controls the world supply of rare earth minerals. According to a US report, over 40% of the semiconductors that sustain the Department of Defense weapons systems and associated infrastructure are imported from China. During the period of 2005-2020, the number of Chinese suppliers in the US defense-industrial supply chain has increased 400% and in 2014-2022, American dependence on Chinese electronics increased by 600 percent. US Ford-class aircraft carriers depend on over 6,500 China-sourced semiconductors to operate. Hence, the US needs China more than China needs the US in trade. It is unprecedented in the history of mankind that China could achieve such phenomenal economic success in a span of 70 years. The stark picture of a devastated China in 1949 compared with the 2025 picture of an ultra modernistic China is unimaginable.

In 1978, China's GDP per capita income was US$229. By 2024, the figure has increased to US $13,136. The IMF published a 2024 estimated figure of $35.2 trillion current international dollar (CID) for China GDP, PPP (Purchasing Power Parity), $28.7 trillion CID for the USA, $14.5 trillion for India and $4.0 trillion for the UK. China's GDP in PPP accounts for nearly 19% of the global economy. Every year, China has a trade surplus with many countries, including the US. The political power of China rises in tandem with its extensive economic power. China is the main leader in the world supply chain. Therefore, many countries, including the US, Japan, India and EU cannot decouple from China . Many Indian pharmaceutical drugs are made with active pharma ingredients imported from China. China has control over many strategic minerals that are required by the American industries. With modern farming techniques and machinery, China's food production output increased exponentially. China's every 5-year plan promotes Chinese economic development progressively. It is predicted that China's economy would surpass the US by 2030.

The US is now beset by many domestic problems of high inflation, interest rate, national debt, unemployment, and government expenditure. In 2024, the US government owes creditors US$1 trillion in interest payment, which is higher than their defense budget of US$841 billion. In 2024 the US national debt reached more than US$35 trillion which is higher than its GDP, and this figure continues to rise. The US substantial financial support for wars in Ukraine and Israel have taken a toll on the US defense budget.

3.3 CHINA TECHNOLOGICAL AND SOCIAL PROGRESS IN 40 YEARS

In the past 40 years, through the successive leaders in the PRC government, China has progressed economically and socially. China has built 45000 km of high speed rail, a network of expressways, new airports, seaports with autonomous port operation, upgrading of public facilities and infrastructure. It has a convenient public transport system with electric buses and an underground subway system with a standard even more advanced than many western countries. The streets are clean and there are many new public gardens and parks built for people to enjoy a stroll, exercise and social dancing. China is well known as being one of the safest places in the world, when people can walk at night in the streets without fear of being attacked. Autonomous bus and robotaxi services are available on the street to reduce traffic jams and accidents. Face recognition is used at the airport for security checks, thereby decreasing labor costs and increasing operational efficiency. Autonomous farm machines are used to improve production and reduce costs. AI is used extensively in agricultural and livestock production. Many cities are converting to smart city standards.

The public enjoys a high standard of medical care at affordable costs in the public hospitals. Although China has a Communist Party government, the country practices socialism with Chinese characteristics. Capitalism is evident in daily life. In 2024, there will be 814 billionaires in China, which is more than the US with 800. China does not practice communism. Communist Party of China (CPC) is just a registered name for a political party. Many European countries also have communist parties. China is very advanced in digital payment via Alipay and WeChat Pay, using a smartphone for payment of purchases of goods and services. Such digital payment using the mobile phone enables Chinese people immense convenience in daily lives and time saving. Also, they deter over-spending by consumers. Using a digital payment system provides easy payment for taxi rides and takeout food at night. The people do not need to go to the bank to withdraw cash, and the banks also save a lot of manpower. Such convenience is really a gift to millions of restaurants and street vendors in the whole country. They do not need to go to the bank to exchange coins. Technology has helped China improve its standard of living, which is better than some western countries. AI and big data are the future engines that propel China development to greater heights.

Farmers in the rural areas use digital payment for their daily business transactions without the need to go to the bank in the city for transaction settlements as in the past, thereby saving time and expenses. China leads the world as a cashless society, promoting a greener environment, eliminating the use of paper without the need for printing and distributing paper money. China is a huge country with 1.4 billion people and hence can save a lot of trees without the need of making paper money, as well as the need of minting coins. The people in the US still use credit cards to pay for goods and service purchases. Using a credit card could lead to over-spending and unnecessary debt. In the US, credit card debt had reached US$ 1 trillion. In Australia, using a credit card for payment incurs a fee. Some vendors in Australia only accept cash payment.

Despite being banned by the US government on chip export, Huawei was able to produce the new Mate 60 Pro mobile phone using in-house production of 7 nanometer chip in August 2023. This smartphone received overwhelming support from users around the world as it supports satellite network communications. The Mate 60

series smartphone supports satellite call functions and short message ,
sending and receiving functions through the BeiDou system without
internet connection. This satellite call function feature is not yet
available in other phone models made by other manufacturers. Huawei
phones are banned in the US for security reasons, claiming the phone
is used to spy on American users. However, iPhones made in China
can be used in the US and China. BYD car model Yangwang U8 is a
luxury plug-in hybrid SUV capable of floating in the event of heavy
flooding. The speed of China's rapid social and technical progress over
the past 40 years has surprised every country. Such speed of industrial,
technical, economic and social progress in China is unprecedented in
human history. It is called "China Speed".

3.4 CHINA TO BUILD A SMART NATION

In order to challenge the threat of US economic hegemonic power, China is to embark on building a smart nation, including the construction of 200 new smart cities to accommodate China's remaining population of 600 million rural residents. China is not short of financial and technical resources to build smart, modern cities. Currently, China invests a lot of money in the Belt and Road Initiatives in many countries to improve their economy and social development. Hence, there is no reason the Chinese government would not invest money to improve the living conditions of the rural farmers and to further improve the country's economy. Every new smart city will provide the children of the farmers free education up to university level that has not been available in the rural areas in the past. The farmers will receive appropriate vocational training for various skill sets. Once they are trained, they can get good salary jobs working in various industries. Their economic contribution will help increase the country's GDP significantly, making China the largest economy in the world. More money will circulate within the country's economy, and hence more jobs are created. Poverty in China will disappear completely. Investment in the new 200 smart cities can reap astronomical economic benefits for years.

As a smart nation, China will incorporate the latest digital technology to improve the country's political, economic and social development. Every provincial government will streamline its administration using IoT and big data digital technology, reducing red tapes and duplication of administrative functions. The education system will apply the latest technology of robotic teaching, thereby reducing the workload of manual teaching staff and improving teaching efficiency. The national school syllabus using AI will be based on the needs of the industries in the country, incorporating feedback from various industries and the teaching staff. As the world advances towards digitalization, quantum computing , AI and big data, the government will emphasize on increasing the number of STEM graduates from the universities. According to a report, by 2025, Chinese universities will produce more than 77,000 STEM PhD graduates per year compared to approximately 40,000 in the United States. If international students are excluded from the US count, Chinese STEM PhD graduates would outnumber their US counterparts more than three-to-one.

China's introduction of digital currency electronic payment (DCEP) will eliminate corruption and fraudulent money transactions. This DCEP is a digital form of legal tender issued by the People's Bank of China to replace paper money. The DCEP application does not need to use the Internet. In the future, digital technologies reshape the global security, industrial and economic landscapes. According to a report, six Chinese universities occupy the top 10 universities in the world. In this ever-changing world influenced by the intense US-China geopolitical rivalry, China is keeping pace with the times and making reforms in various political, economic and social systems to suit the prevailing world order. China is now going through a transformation of industrial transition from the low to high-end products in the value chain.

3.5 China's Foreign Trade

In 2024, China's import and export value exceeded 43.85 trillion yuan ($5.98 trillion), maintaining its position as the world's largest trading nation in goods for six consecutive years. In 2022, China's imports and exports to ASEAN, the EU, and the US reached 6.52 trillion yuan ($945 billion), 5.65 trillion yuan ($818 billion) and 5.05 trillion yuan ($732 billion) respectively. In the same period, imports and exports to countries along the "Belt and Road" increased by 19.4%, accounting for 32.9% of China's total foreign trade, and imports and exports to other RCEP (Regional Comprehensive Economic Partnership) member states increased by 7.5%. In addition, imports of crude oil, natural gas, coal, and other energy products totaled 3.19 trillion yuan, ($462 billion) accounting for 17.6 percent of the total import value. Agricultural imports reached 1.57 trillion yuan ($227 billion), accounting for 8.7% of the total value of imports. In 2023 China spent US$ 64.3 billion on LNG imports of 120,000,000 tons.

Despite many sanctions and high tariffs, China exports to the US in July 2024 reached a record high. The US introduces the National Defense Authorization Act (FY25 NDAA) banning DJI drone sales in the US. China controls over 70% of the world's drone market. The US police department uses DJI drones to find missing children and to capture violent criminals. The police are unable to use US made drones to fly to high mountains to locate missing hikers. Only DJI drones can do the job. The Secret Service Agency uses DJI drones to do aerial surveillance. DJI drones are used by 90% of American search and rescue teams. Pentagon gets a sanction waiver from the government to buy Huawei telecom gear and DJI drone.

China's GDP in 2022 reached 121.02 trillion yuan ($17.54 trillion), an increase of 3.0% over the previous year. According to the World Bank's measure, China's 2021 gross domestic product has surpassed that of the US in terms of purchasing power parity. On an annual basis, China accounts for a third of total global economic growth. Therefore, China does not need to worry about US and China bilateral trade derailing because China's internal and external markets are huge. China is the world's factory, and is also the leader of the global supply chain. China has bilateral trade with 120 countries in the world. China's global multilateral trade has established itself as the global trade engine driver. So the US cannot cut China out of the global economy. Donald Trump in 2025 threatens every country to impose hefty tariffs on them if they do not curtail trade with China. Many countries ignore his threat.

As time passes by and the world is undergoing uncertain changes, changes in the foreign policy of the US and its economic downturn will affect the geopolitical situation and social changes around the world. China needs to update its diplomatic, political, trade, technological, and financial policies and guidelines to adapt to the prevailing global environment. Despite protest from the US, Germany in 2024 is doubling down on investments in China. The German Bank Deutsche Bundesbank reports German corporations, mainly carmakers, invested 2.48 billion Euro in China during the first 3 months in 2024 and in the next quarter invested further 4.8 billion Euro.

Some economists estimate that by 2030, the Asia-Pacific region will account for 40 percent of global GDP and two-thirds of the world's middle class, while China's middle class will grow from 300 million to 600 million. At that point, the Chinese digital yuan will become a major international trade and settlement currency. BRICS started trading in member currencies, moving away from the US dollar and SWIFT transact system which involves 2%-3% double foreign exchange fees from local currency to US dollar and vice versa. Many countries have applied to join the BRICS. The economic power of BRICS will gradually expand, surpassing the G7 group.

3.6 THE US ACTIVELY RESTRAINS CHINA'S ECONOMIC DEVELOPMENT

After WWII, a bipolar world was created by the USA and the Soviet Union. Both were two superpowers with strong power in the global economy, culture, international politics and military. These two superpowers tried to dominate each other for decades. After the Soviet Union was formally dissolved as a sovereign state on 26 December 1991, the US emerged as the sole superpower. A new unipolar world order emerges and is dominated by the US . The unipolar world has lasted till 2012 when China ascended to the world's second-largest economy status.

A new multipolar world order has emerged, with China and Russia to eclipse the sole superpower of the US. China and Russia power which includes their alliance in BRICS and SCO possess immense geopolitical and economic power, with rich natural resources in oil and minerals. China, Russia, and their allies combined power in population, GDP, military, and economy is larger than that of the US. Their allies can help the two powers to achieve their goals. The US has been employing the following measures to restrain China's economic development: namely bribery, propaganda, military threat and trade war.

- The CIA has a $1.6 billion budget to bribe agents around the world to cause civil unrest in China. The CIA overtly supported the worst riot in Hong Kong history in 2019-2020. Each of the thousands of young demonstrators waving American flags was paid by the CIA. The activist leaders Jimmy Lai, Martin Lee, Janet Pang and Joshua Wong in 2019 even went to the US, addressing the Congress to seek US help to topple the Hong

Kong government. The young rioters also tried to paralyze Hong Kong International Airport by attacking foreign travelers. The CIA also incited confrontation between China and India, and between the Philippines and China. The CIA also engaged the Dalai Lama to cause civil unrest in Tibet and its covert operation in Tibet for years. The most obvious CIA covert operation is to incite independence in Taiwan and sell expensive weapons to the independence political party government. The US bribed overseas Xinjiang activists to write fake stories about Chinese government carrying out genocide on Uyghur Muslims and hard labor in the cotton fields without evidence.

- The numerous US private media and their foreign affiliates daily issue disinformation on China without human rights, individual freedom and freedom of speech. They broadcast that the CPC is a repressive government towards the citizens, even though the CPC has lifted 800 million people out of poverty. The US politicians still publicized the stale old 1976 Tiananmen student demonstration as human right violation. The CIA propaganda factory will continue to spread the Xinjiang genocide and forced labor disinformation for a long time, as they think spreading lies for a long time would become fact. The American technique of smearing China's image is a very cheap way of toppling a government in this era of internet around the world. Many countries now have an unsavory image of China.

- The US has been employing a fleet of aircraft carriers and other warships patrolling the South China Sea since 1949 . Very often, the US also invited navies from UK, Germany and France to participate in naval exercises to show off the military might of the US and its allies in the South China Sea to threaten China. The

US forms the Five Eyes, Quad, and AUKUS military alliances to encircle China's coastal areas. Lately, the US-led NATO aims to establish a NATO 2.0 in Asia with the objective of selling more weapons to Asian countries. Ex-Sectary of State Anthony Blinken often brought his partner Secretary of Defense Lloyd Austin on his diplomatic trips to Asian countries. Such frequent diplomatic trips involving the defense secretary are rare in normal diplomatic trips. Austin occasionally made foreign affairs statements about China's threat. The CIA propaganda factory has created the China-threat, Russia-threat, Iran-threat, and North-Korea-threat as tools to coerce other countries to buy US weapons for defense. US's weapon manufacturers MIC is a huge conglomerate which has close relations with politicians in the Congress.

- The US government started a trade war with China on June 15, 2018. President Donald Trump released a list of $34 billion worth of Chinese goods to face a 25% tariff．A second list worth $16 billion of Chinese goods was also released later. Tariffs of 30% to 50% were imposed on solar panels and washing machines, and 25% tariffs on steel and 10% on aluminum. He passed a bill to sanction Huawei 5G communication equipment and smartphones. Under the Joe Biden administration, he imposed further sanctions on many Chinese companies in the high-tech industries. He imposed a 100% tariff on Chinese EV imports, 50% on semiconductor and EV batteries, The 25-50% tariff on ship-to-shore cranes, medical syringes and needles, personal protective equipment, respirators and face masks, rubber medical and surgical gloves.

All these tariff increases on Chinese imports will definitely increase the costs of living and inflation, as US taxpayers will have to bear the increase in import tax. Over 40 percent of the semiconductors that operate DoD weapons systems and associated infrastructure are sourced from China. Biden passed a bill banning the import of drones made by Chinese manufacturer DJI. However, the bill was rescinded one month later after many protests from farmers, self-defense force, and emergency services. Even the Pentagon buys DJI drones for military application. The Congress prohibits purchase of Chinese specialty materials used in military weapons. However, the Pentagon got a waiver to buy a Chinese special mineral in the assembly of the F35 jet fighters. How to fight China when the US needs materials from China for special materials like rare earth minerals is a mystery. China has imposed on the US exports of rare earth minerals which are used in manufacturing of military equipment, EV batteries, and solar panels.

How long the US would continue to use bribery , propaganda, military threat and trade war against China is indeterminate as the US is bent on creating trouble for the Chinese economy. Currently, the US dollar is beginning to wane as an international trade and reserve currency. The US is beset with ever rising national debt, high inflation and interest rate and its proxy wars around the world. US politicians in the multipolar world are wasting their precious time doing decoupling or derisking, friendshoring to weaken China economic progress and disrupting the world supply chain. Such negative actions do not bore well for the US economy. China makes up 35% of the total global output for manufacturing and 29% of value-add of world manufacturing.

China trades with 120 countries and is the largest export country in the world. Hence, even if the US decouples from China in trade. China can continue to trade with other countries and expand its products around the world. China via BRICS, SCO and BRI will enhance trade with all other members. China's economy will not collapse just because of the US anti-China policy. In 2023, the total value of the US trade in goods with China amounted to around US$575 billion, comprising US$147.8 billion export value and a US$427.2 billion import value. Many US manufacturers still depend on Chinese imports for the production of their final products. US manufacturing contributes only 10.3% to the US GDP.

3.7 China's Third Economic Revolution

D eng Xiaoping started China's second economic revolution in 1978 with his open market economy reform policy, overhauling Mao Zedong's closed door economic policy. Since 2012 China attained the status of the world's second-largest economy , the US under President Barack Obama felt threatened by the rise of China. He began a new pivot to Asia foreign policy, integrating diplomatic, military and economic strategy with allies in the Asian Pacific regions that changed drastically the US-China relationship. His anti-China policy basically is to cripple China's economy. His policy is similar to the US policy of crippling the Japanese economy by forcing Japan to sign the Plaza Accord in 1985, forcing Japan to appreciate its Japanese Yen against the US dollar. This results in Japanese exports becoming non-competitive. It is also similar to the proxy war in Ukraine that the US wants to cripple the Russian economy. The US is steadfast in wanting to cripple the Chinese economy, even at the expense of their own economy. In addition, the US wants its allies in the West to follow US anti-China economic policy.

The current international geopolitical, trade and economic situation is undergoing major and rapid changes. The US has destroyed the global supply chain that has been established for the past half century and now wants to homeshore American manufacturing from China. In 2024, Saudi Arabia officially ended its 50-year 'Petrodollar' agreement with the US, shifting to a multicurrency system for oil sales. China has started using RMB to purchase oil from oil exporting countries, thus saving substantial cost by avoiding transactions via SWIFT. To avoid paying high US tariffs, many Chinese companies have shifted their manufacturing of products of low value chain to ASEAN countries, which are not affected by US high tariffs.

2024 is the turning point of China's third economic revolution. Amidst the relentless anti-China campaigns encompassing trade, technology, finance, and chip wars, which significantly impact China's exports to the US and EU, China is compelled to transition its economy towards higher value-added sectors, particularly in high-tech industries and Fintech. Although China has made breakthroughs in some technological fields, there is still some distance between China and Western and Japanese technology in areas like precision instruments and machineries and extreme ultraviolet (EUV) lithography machines in the computer chip-making process. The Chinese government will provide substantial financial aid to companies that are involved in designing and manufacturing of high-end semiconductor chips. With more government investment incentives to private enterprises in areas where China faces US sanctions and high tariffs, more innovations in the high-tech areas will appear in the future.

China's electric vehicles are now more than just a car but also a machine for entertainment and communication tools. The car incorporates a smart system to park and retrieve in a car park. Such an AI system is convenient for shoppers to avoid rain, snow, and the problem of carrying a child and groceries together to look for the car in the open car park. The car can automatically reach the mall entrance/exit undercover. As the world's aging population increases, many disabled people have mobility problems of going out of their homes to go shopping, visiting friends or care centers on their own.

Chinese EV manufacturers can design an autonomous EV that allows easy access of the EV for a disabled person on an electric wheelchair.

Such EV can transport the disabled person without the need for a companion anywhere, such as going to a hospital without waiting for an ambulance. Such EV should be popular in the vast land of the US, Canada, Brazil, Argentina, Australia, Saudi Arabia, Russia where many disabled people live in small towns that have little or no public transport or taxi. With such autonomous special EV, this should make their lives more convenient and comfortable for shopping and visiting, especially during winter or inclement weather. China can design an affordable EV as a short term solution to the housing crisis in many Western countries and America. The fully furnished EV can provide shelter and mobility for a young family of three comfortably at a price of less than US$100,000.

To avoid intense geopolitical trade disputes with the US and Europe, there are many new export industries that can be developed by Chinese entrepreneurs for new emerging markets in Africa, Central Asia, South East Asia and South America. China can export to Third World countries mobile medical clinic vehicles equipped with appropriate medical devices, instruments, lab, medical equipment for patients in small towns and villages without a hospital. China can offer a turnkey project for building a small scale hospital in small cities or towns in various countries. The hospital will be furnished with all the necessary medical facilities for a small population.

Using 5G communication and the internet, Chinese doctors can provide remote medical assistance to any hospital in foreign countries. With the latest language translation using smartphones, there is no problem in communication between Chinese and other nationalities. China can also offer training to foreign students to become healthcare service providers like nursing and technicians to maintain hospital equipment and facilities. Such medical assistance will enhance China's soft image. Many Third World countries suffer from food shortages. China can help set up turnkey projects on AI fish farming that is suitable for local consumption to satisfy demands. There is an immense opportunity in China for entrepreneurs to apply digital technology in many new industries on factory productivity, fintech, traffic systems, sanitary systems, AI aided agriculture industries, water conservation and land reclamation in desert regions.

Africa has a population of about 1.54 billion people. Many countries are still in the developmental stage. They have limited public transportation infrastructure, making the common bicycles a practical and convenient mode of transportation. Bicycles are also used for food delivery and transportation of goods. Hence, there is a business opportunity for Chinese investors to produce the ubiquitous bicycle in African countries. Other vehicles that are needed in Africa are an electric tricycle for food, cargo, and passenger carriage and electric bicycle. The electric tricycle trucks are convenient for farmers to transport their produce to the city markets. Chinese investors can install a central electric charger at the village using solar panels and storage batteries.

Chinese smartphone companies can set up factories in African countries to produce basic smartphones with fewer functions than those sold in China. This should reduce the cost of production and the sale price of the phone, making it more affordable to the local population. China can set up factories in Africa to manufacture low cost household items to meet the local requirements. These low-cost industries can be upgraded in tandem with economic progress in each country. Many raw materials and components can be imported from China. Such investments follow the copybook of China's Deng Xiaoping's open market economy. All the new projects can be implemented under the BRI programs.

The US, Germany and Japan, have set up factories around the world to consolidate their economic power in the past. Australia car markets had been dominated by the Japanese car manufacturers after forcing out the American car manufacturers. Japanese TV, electronic products factories are ubiquitous in South East Asian countries and from there, the final products are exported to other countries. The Chinese government needs to encourage and assist private companies to follow the footsteps of these advanced countries to set up factories around the world. There are many middle-income families in ASEAN countries, Africa, Central Asia, the Middle East, and South America. With China's massive supply chain, the supply of components and materials is not an issue to support overseas factories making household items, commercial products, medicines, and transport vehicles and machinery.

With BRICS, SCO and BRI, China should have no problems setting up factories in the member country. Members will use domestic currencies for cross-border trade which exclude the US, Japan, EU, and South Korea. Hence, these member countries provide a huge market for products made by Chinese factories located in the respective member countries. This is the opportune time for China to embark on exports made in foreign countries. Soon, goods made by China will float the world market. China's investment in 200 new smart cities to accommodate 600 million rural residents and the establishment of an artificial intelligence agriculture industry will boost China's economic progress significantly.

US decoupling of trade from China will have little impact on the Chinese economy. As long as China focuses on economic development and global expansion to enhance its economic power, by 2030 China will surpass the US to become the world's largest economy and highest GDP. War with the US is not China's foreign policy. Even if China can win the war , it is of no use to China, as the war will have a significant impact on China and the world economy. Creating an enemy with the US serves no useful purpose to China. In any case, the US navy cannot fight China in the South China Sea to win as the US has conducted 10 war games and the US has never won one game against China. Time is on China's side, awaiting the US empire to crumble without China firing a bullet. The US can continue to prosper only if it can work with and not against China.

3.8 CHINA UPDATE NATIONAL ECONOMY

With the launch of China's dual circulation economy, an average annual economic growth rate of 5%-6%, and the expanding 'Belt and Road' project, China aims to avoid a hot war with the US, thereby enabling it to upgrade its economy and become the world's largest as soon as possible. The US has declared that it will stop China's economic progress and start implementing various trade wars, science and technology wars, culture wars, diplomatic containment wars, misinformation wars etc. against China. In the past, China has experienced threats from its borders. China built the Great Wall to keep China safe. Now China needs to build a great economic wall to resist the US threat. China is the world's largest market and is expected to import goods worth more than US$22 trillion in the next decade, which will inject immediate and lasting vitality into the world economy. China encourages the exporting countries to use the digital yuan for settlement and avoid using the US dollar.

The Chinese government has realized that under the dual-circular economic development plan, China will no longer target trade surplus and over-dependency on expanding exports. The expansion of imports will give China an advantage in future trade negotiations, promoting the internationalization of the digital yuan. China will import low value products from other countries and then export high value products. This product exchange system will accelerate the development of high value-added products in China.

The use of an international digital yuan could lead to rapid growth of China's per capita GDP. China has emphasized that achieving scientific and technological independence helps improve its innovation capacity. From now on, Chinese factories producing low-value products will move to low-cost countries in Southeast and South Asia. These low-value products will be exported to earn foreign exchange and will help the economies of these countries. China would benefit from outsourcing the manufacturing of these low-value products overseas. Chinese workers' wages are rising, and the manufacturing of low-value goods is no longer viable.

The world has undergone profound changes unseen in a century. China maintains a strategic focus on continuing its opening-up and reform, adhering to a dual circulation plan. This approach aims to attract foreign investment, increase employment, and foster competition to stimulate the production of high-end products. Furthermore, it seeks to draw in new ideas and technologies, energize economic and financial market activities, and promote the rapid turnover of the international digital yuan. China realizes competition does improve quality and reduce manufacturing costs. Modern products have shorter life spans. With so many projects to implement and on such a broad scale, China is desperate in need of talents.

China must relax its immigration policy, emulate US immigration policy, and let talented people from abroad to work in China. As Ren Zhengfei wants to increase Huawei progress and to overcome various sanctions imposed by the US, his company pays top dollar to hire foreign talents to create new products for Huawei to sell worldwide. Huawei is a multinational company employing local talents in every branch abroad. Huawei is a private company with assets worth US$178 billion and an operating income of US$14.74 billion in 2023. His company's profits are shared amongst his employees. Huawei has 207,000 employees and operates in over 170 countries and regions.

His responsibilities towards all his employees are monumental and arduous. His affordable products must always be one step ahead of his competitors worldwide.

The government appreciates that a private company like ByteDance can secure worldwide acceptance of its TikTok Apps around the world. It is one of the world's most popular social media platforms. In April 2020, TikTok surpassed two billion mobile downloads worldwide. In September 2021, TikTok reported that it had reached 1 billion users. In 2021, TikTok earned $4 billion in advertising revenue and is expected to generate $14.15 billion in revenue in 2023. The US threatens ByteDance to sell TikTok to a US company, otherwise it will be banned.

3.9 Effect of Ukraine War on China Economy

The Russia-Ukraine war started on 24 February 2022 has generated new challenges and new problems in international politics and trade. Ukraine's major export of wheat to the world market is abruptly affected . So are the major Russian exports of fertilizer, oil and natural gas. The war has disrupted the world supply chain and price. The contest in the ideological field is very complicated, affecting China's economic development. The proxy war in Ukraine created by NATO led by the US has one sole purpose of crippling the Russian economy forever. The US coerces Ukraine President Zelensky to apply for NATO membership, which is fiercely opposed by Russia, as NATO can station troops and weapons on the Russian border. The war involved a massive supply of weapons from the US and NATO members to Ukraine. The US also rallies other countries to sanction Russia on trade.

The US is determined to win the proxy war, which can be considered as a mini-WW III. Till to date, there is no sign of abating as EU/NATO countries are not willing to agree to a ceasefire. In fact, the war has actually intensified by the Ukrainian military beginning to massively attack deep into Russian territories with US military aids. Denmark offers Ukraine 19 F16 jets and Norway offers six F16 jets. Thus far, estimates of Ukrainian soldiers killed and wounded are 350,000-500,000 and Russian soldiers killed and wounded as 200,000-700,000. Ukrainian civilian casualties are estimated at 35,000 killed and wounded. Many mercenary soldiers are also killed. There are 3.7 million people displaced within Ukraine and 6.3 million Ukrainians have fled the war as refugees and asylum-seekers in Europe and other countries.

According to a 2024 US media report, the leading 15 American defense contractors are forecast to gain a cash flow of US$52 billion in 2026 for massive weapon orders. The 2024 US Biden administration aid bill for Ukraine, Taiwan, and Israel is allocated nearly US$13 billion for weapons production at America's five defense contractors. The UK government committed £7.6 billion for military aid to Ukraine. The war machine is a cash cow for MIC. The American MIC has so much cash flowing into their pocket that the only way to spend the cash is to pay dividends to their shareholders, which include many politicians. Hence, the Ukrainian war is expected to continue for a long time, and the people in Ukraine will continue to suffer. The country may end up looking like Afghanistan. The whole country is ruined and a million people die and are wounded. What is the Ukrainian purpose of fighting a losing battle for another country?

After the undersea gas pipeline was severely damaged by U.S. agents, thereby cutting off Russia's cheap gas supply to Germany and other EU countries, as well as sanctions on Russian oil exports to the EU, the economies in Germany and the EU were badly affected. Inflation has soared and the cost of living has shot up. Every European and American company retreated from Russia. No bank is allowed to transact US dollars with Russian banks. Suddenly, the Chinese RMB became the favorite currency for trade in Russia. Economic integration between Russia and China has accelerated dramatically, with total trade between them reaching US$240 billion in 2023. The volume of Russian crude shipped to China jumped 24% in 2023 to 107.02 million metric tons, valued at US$60.64 billion. Chinese EV swarms the Russian car market when European and Japanese cars are withdrawn from Russia.

Russia's GDP continues to grow in 2023, as it expands by 3.6%. Such a situation is not what the US expected. Russia depends on export of oil, gas, wheat, minerals, and coal for its income. The US proxy war in Ukraine has further disrupted the world supply chain. The geopolitical conflict between China and the US is further complicated by the war in Ukraine and in Israel, and the world is split between the West led by the US and the Global South led by China. China's political influence is enhanced through initiatives like BRI, SCO, BRICS, AIIB, and RCEP on the world stage, while the US image is tarnished by its extensive use of sanctions against many countries and the threat of confiscating their US assets held in American banks.

3.10 US Maintains Steadfast Policy to Cripple Chinese Economy

At present, American politicians do not feel right about the international supply chain contributing towards China's rapid economic progress. They want to disrupt China's supply chain in the hope that it will stop China's economic development. Trump's harsh foreign policy toward China has hurt bilateral trade relations and supply chains, as the US has imposed sanctions on certain Chinese exports and raised tariffs on Chinese imports. He has banned American chip exports to China and blocked its Allies from using Huawei 5G communication equipment. His actions have also disrupted the functioning of world supply chains. America not only spreads misinformation about China and interferes with China's internal affairs every day, but also wants its allies to follow US anti-China foreign policy that could lead to the detriment of their own economy.

The US has succeeded in convincing Germany, the UK, and other Western countries to dislodge their Huawei 5G communication equipment from their networks, thereby forcing Dutch company ASML not to sell its most advanced extreme ultraviolet technology to China and causing ASML to lose billions worth of sales each year. Similar actions also applied to Japanese and Korean chipmakers not to sell chip making equipment to China. Many mid-stream American technology hardware firms have started moving out of China to India and Vietnam to avoid US government ban on chip hardware to China.

China now exports more to emerging markets than it does to developed markets. China realizes that if its economy depends too much on the US and if the US begins to sanction Chinese companies' products , China will be strangled and many workers will lose their jobs, causing havoc in social stability. Hence, China introduces dual circulation economic policy to encourage domestic consumption. With BRICS de-dollarization and BRI projects performed in the emerging countries, China has spread its influence and cross-border trades, absorbing Chinese products at affordable prices.

Meanwhile, China controls the production of rare earth elements that are much needed by US and Western countries for their industries, especially high-tech consumer products such as mobile phones, solar panels, wind turbines, computer hard drives, electric vehicles, flat screen monitors, and TV. They are also used in defense applications including electronic displays, guidance systems, magnets, lasers, radar, and sonar systems. These elements are important in the manufacturing of modern fighter jets. China has imposed an export ban on several rare earth elements.

China adopts a neutral attitude towards US hegemonic actions on its inhumane war in Ukraine and Palestine. Although the US warships have been patrolling the South China Sea since 1949 with occasional naval exercises with its allies, China remains steadfast not to be provoked into a military conflict with the US. There is no benefit for China to start a war with the US unless the US begins to attack the Chinese mainland and destroy Chinese infrastructures and military installations. The US has enough naval power to block Chinese merchant ships passing through the Strait of Malacca, even for a few days. The consequence of such blockage can be very unsettling in the world of geopolitics and tension amongst Asian countries.

Military conflict in the Strait of Malacca also affects ships of other countries from Europe and Asia. Consequently, their imports and exports operations will be impacted. 80% of China's Middle East oil shipment passes through the Strait of Malacca. Any international military conflict will jeopardize domestic and foreign investments. One way to counter the US threat to block the Strait of Malacca is for China and ASEAN countries to form a military alliance to protect free maritime passage of the Strait for all countries under the auspices of the military alliance. However, ASEAN countries are not keen on military alliances to avoid antagonizing the US. If the US starts to create maritime conflicts with China at Malacca Straits, other countries' trade will be affected, and they will protest against the US.

Russia now accounts for 19% of China's oil imports, while Saudi Arabia makes up 15%. The volume of Russian crude shipped to China jumped 24% in 2023 to 107.02 million metric tons, valued at US$60.64 billion. Much of the oil is shipped via the Malacca Strait. China also imports oil from Saudi Arabia, the US, and Brazil. In 2023, China spent $337.5 billion on oil imports. If China can invest the same amount of US$337.5 billion yearly on renewable energy projects to replace oil imports, this investment will generate enormous benefits to the country's economy and employment. The cumulative annual investment money of $337.5 billion will be circulated for years within the country to generate more economic benefits.

3.11 Shenzhen's Economic Progress, a Template for Other Chinese Cities

Shenzhen was a small town with about 30000 residents before it was selected by Deng Xiaoping in 1978 as a pioneer Special Economic Zone for his famous open market economy policy. He encouraged people to be entrepreneurs. His many quotes reverberated throughout the country, for examples: "cross the river by touching the stones", "poverty is not socialism," "to be rich is glorious," "we must not fear to adopt the advanced management methods applied in capitalist countries," "the very essence of socialism is the liberation and development of the productive systems," and "socialism and market economy are not incompatible." In essence, he relinquished communism and adopted capitalism. In a span of 30 years, Shenzhen has a GDP larger than that of Greece and has more billionaires than any American city.

Shenzhen industries started with factories contracted to manufacture for foreign brands with low skill laborers. The factories made affordable clothings, toys and electronic goods sold worldwide. Initially, many investors came from Hong Kong, Taiwan, and South East Asia. Gradually, Shenzhen attracted many Western companies that contracted to local companies to manufacture high value products in the supply chain. Finally, Shenzhen upgrades itself to become China's second Silicon Valley and is also seen as the technology capital of the world because of its combination of software and hardware capabilities that lead to prosperous economic growth. It is said that 90% of the world's electronic products are manufactured in Shenzhen. With tens of thousands of factories, 5,000 product integrators and thousands of design firms, the city has become a one-stop shop for integrated circuits, chips, LED lights and touch screens.

Shenzhen rapid industrial growth depends on availability of skilled workers and science and engineering graduates. Shenzhen economy has been helped greatly by being close to Hong Kong financial hub and international business connection. Deng Xiaoping had the foresight to select the small town Shenzhen for starting his open market economy policy and not other mature cities like Beijing, Shanghai, Guangzhou, or Nanjing. Shenzhen has no historical baggage to hinter its social and political advancement. It accepts Western modern ideas readily. Shenzhen is home to 20 percent of China's PhDs, has the highest percentage of entrepreneurs in the country, and has produced more billionaires than anywhere else in China.

Shenzhen is dominated by high-tech industries and advanced technologies. Shenzhen has many industries with high value-add, low energy consumption and low carbon footprint. More than half of China's 5G technology research and development and innovation applications originate from Shenzhen, and the upstream and downstream of the industrial chain are interconnected, which is very convenient for the manufacturing industry. The Shenzhen government encourages more high-tech developments involved in the new generation of information technology.

In 2020, Shenzhen becomes the first major Chinese city in terms of industrial value-add, surpassing Shanghai. In October 2020, the central government selected Shenzhen as the first test city, selecting 50,000 residents to apply and be issued with digital currency worth 10 million yuan. The government sends money through mobile "red envelopes" that allow them to pay for purchases in digital yuan at selected companies. Shenzhen is the first major city in China and the world to operate electric buses. The city benefits from the absence of air pollution from diesel engines.

With a population of 17.5 million in 2020, Shenzhen is the third most populous city by urban population in China, after Shanghai and Beijing. The Port of Shenzhen is the world's fourth-busiest container port. In thirty years, the city's economy and population boomed and has since emerged as a hub for technology, international trade and finance. It is the home to the Shenzhen Stock Exchange, one of the largest stock exchanges in the world by market capitalization. Its nominal GDP, which is larger than Greece, has surpassed those of its neighboring cities of Guangzhou and Hong Kong and is now among those of the cities with the ten largest economies in the world.

Shenzhen also is the eighth most competitive and largest financial center in the world, and has 10 firms making the Fortune Global 500 list. It has the second-largest number of skyscrapers of any city in the world, the 19th largest scientific research output of any city in the world, and several notable educational institutions, such as Shenzhen University, Southern University of Science and Technology, and Shenzhen Technology University. China's large technology corporations, such as Huawei, Tencent, and DJI are located in Shenzhen. Shenzhen is China's leading innovation hub, with a GDP of nearly US$500 billion, over 40% of which is generated by high-tech industries. According to a blueprint released by Shenzhen's Development and Reform Commission, the city hopes to create a number of "unicorn" companies in biotechnology, semiconductors, quantum technology and other emerging fields.

The prosperity of Shenzhen is not controlled by the central government, but depends on the ability of Shenzhen residents and its local government. Other cities may wish to learn from Shenzhen people's dream of starting a business and becoming rich. Shenzhen, Hong Kong, Singapore, and San Francisco are all big cities of immigrants with dreams of starting businesses and becoming rich. New immigrants have a new diversity of ideas and are not bound by traditional conservative habits. The city of Shenzhen looks more like a capitalist city in the West.

Shenzhen is indeed a self-made city, with freedom of being a successful entrepreneur, dare to take risks and cross the river by touching the stones. To get the talents he needs , Huawei boss Ren Zhengfei is willing to pay the right talent a one million yuan salary annually. The right talent can contribute to Huawei more than a million yuan worth annually. In 2024, Huawei spent 179.7 billion yuan on research and development, representing 20.8% of the total revenue of the company. Such high R&D expenditure is very rare in industries, either private or public.

3.12 China Needs Talents to Accelerate its Economic Development

Over the years, the most prominent problem in China's scientific and technological development is the problem of shortage of talents. Despite its huge population, China has a severe shortage of world-class experts in science and technology. According to statistics, the Nobel Prize has been continued for 119 years, and the US leads the way, with 260 Nobel Prize winners, accounting for 40% of the total number. The most important reason the US has so many Nobel Prizes winners was that it attracts the best talent from all over the world to immigrate to the US to work and live. The US has a world-class policy of sourcing talents from the world.

The US has a free environment for people to do research. It is also the country with the strongest original innovation ability and the development of emerging industries in information, network, biology, new energy, nanomaterials and so on. The US has many world-renowned universities, cultivating outstanding talents. There are many angel investment groups in the US that provide financial support for technology and science entrepreneurs. Can this kind of free and open environment for study, research and development, and entrepreneurship in American universities be replicated in China? It may be difficult because of the differences in culture and teaching methods in China.

3.13 CHINA NEW SYSTEM INTEGRATION INDUSTRIES

Industrial integration is directly related to the concentration, specialization, and cooperation of production and supply chain, and promotes the improvement of efficiency. A typical system integration industry is commercial aircraft. It takes 6 million parts to build a Boeing 747 airplane. Adding hundreds of accessories, such as tires, engines, oxygen systems, drinking systems, waste removal systems, cockpit instruments, and more. Boeing assembles all the parts and accessories into an airplane. Many parts and accessories are imported. Therefore, the assembly of an aircraft for immediate delivery requires the immediate cooperation of a team of suppliers to produce the lowest cost of an aircraft. Every part and accessory is designed and manufactured by a specialized Original Equipment Manufacturer (OEM). This is because Boeing cannot design and manufacture every component for aircraft assembly, and even if it could, the resulting manufacturing costs would be too high to remain competitive. Apple's iPhone is also a result of parts integration. All mobile phone accessories are made and supplied by other manufacturers. Apple concentrates on the design and outsources everything else. Both Boeing aircraft and Apple iPhone are products of highly integrated systems of supply chain.

The marine transport business is a systems integration industry. Freight forwarder companies combine logistics, packaging, warehousing, storage, cargo customs procedures, insurance, trucking and door-to-door delivery. Hence, marine transport business is a supply chain operation, missing one link will result in an interruption of the entire shipping operation. In late 2021, a global shortage of container shipments and a shortage of truck drivers at western US ports caused many factories to suspend operations. The supply chain of a globally integrated business cannot be interrupted in any way for it to function efficiently. According to the US, "Maritime News" reported that the global shipping container inventory in 2022 is 44.2 million TEUs, 95% of which is made in China.

In 2020, 96 of the 500 largest global companies (almost one-fifth) were Chinese SOEs, accounting for more than US$63 trillion in combined assets.14 Fortune Global 500 includes Chinese national companies in the shipping industry, such as CMG and COSCO, both of which are shareholders of CIMC, a major Chinese shipping container producer. This also includes Chinese steel companies providing steel inputs to producers in the shipping industry, such as China Baowu Steel Group, HBIS Group, and Ansteel Group, among others. It is an impressive economic achievement in a span of 74 years since PRC was established in 1949 without a name in the Fortune 500. The world's largest producer of intermodal shipping containers, CIMC, is a key link in a chain of shipping-related Chinese companies owned by China's State-owned assets. CIMC became the first enterprise with sales volume, breaking two million TEUs, in the global container industry.

According to a report in 2024, China already accounts for 35% of global industrial production and 90% of global shipbuilding orders. In 2023, China received more than 1500 orders for new ships and the US had only 5 new ships ordered. China trades with 140 countries around the world, and therefore it is natural for China to have most ships transporting exports to various countries in the world. However, China does not own the largest shipping company in the world. This discrepancy should be corrected so that China will not be affected by sanctions imposed by the US and its allies.

During the COVID-19 crisis, many shipping operations around the world were affected by a shortage of containers (TEU) as ships have been stranded at many ports. Chinese shipping companies engaged in the export business were the most affected, as containers were stranded in overseas ports, especially in the US west coast ports. These stranded TEUs were not returned to the original port in China for loading. Many freight forwarders had freight stuck in warehouses. The same container shortage also occurred in March 2021. When a container ship, the Ever Given, ran aground in the Suez Canal, blocking it for six days. The ship was then towed to another port for repairs and could not be unloaded for several weeks, resulting in tens of thousands of containers delayed return to the original port of loading. The lack of containers affects global freight transport.

China should set up the world's largest container leasing company, possessing the world's largest number of containers. The company can use the latest Internet of Things (IoT) and blockchain to track every one of its containers around the world. It will have container data for all seaborne needs at each port and support its customers by setting up several regional centers to keep spare containers for emergency needs. Empty containers can be shipped by the company's charter ship to any port that meets customer needs.

China is the world's largest trading country and hence should have the world's largest container leasing company. This container leasing company should act like an insurance company that helps cure a container shortage problem for a shipping company. A delay of shipping on time due to container shortage could lead to the customer losing millions of lost business opportunities. China operates or has ownership of 115 ports around the world. The shipping business will grow rapidly with African countries joining BRICS and the 2024 conference of China-Africa The summit ended in Beijing in September 2024 with China promising to offer US$ 51 billion for African countries' economic projects. Shipping is an evergreen industry.

During the COVID-19 pandemic in Wuhan, the Chinese government achieved the world's fastest hospital building record. A 60,000-square-metre structure with a space for 1,000 beds and 30 intensive care wards built in just 10 days. International hospital design and construction turnkey projects is a very good multinational industry. China could start a turnkey engineering business and build a fully equipped hospital with beds, ICU, X-ray machines, CT scans, laboratory , operating rooms, wards, hospital equipment and instruments, hospital kitchens and cold rooms, laundry rooms, garbage disposal systems. Hospitals will be equipped with air conditioning systems that filter bacteria and emergency power supplies. The hospital administration has a comprehensive computer system for patient admission procedures, patient records, medical history and hospital charges, and is linked to the health insurance companies, and patient admission procedures.

Turnkey engineering involves the design and construction of hospitals of varying sizes that can accommodate hundreds to thousands of patients. Training will be provided for the use and operation of all the equipment and instruments, maintenance contracts for hospital computer systems, equipment, and instruments. This turnkey hospital project is an evergreen industry. It will be the first of its kind in the world. The project is suited to countries in Africa, Middle East , Central Asia, South Asia and South America. Mobile clinics on a mobile vehicle can also be provided as an ancillary project. The turnkey project can provide a fully autonomous ambulance to transport patients to the hospital. Remote surgery and medical consultation can also be provided via 5G satellite communication with hospitals in China and using automatic translation to local language.

3.14 CHINA'S ENTERTAINMENT INDUSTRIES

A country's culture has a huge impact on its economic and social development and on its international image. Hollywood films, TV, VOD, and streaming have influenced other countries around the world to imitate American culture, and it also gives people around the world a positive impression of the US. Hollywood is undoubtedly a powerful tool for American soft power on the international stage. Hollywood films not only unite a nation around the values and geopolitical representations of the US but also the entire world. The Hollywood film industries are famous for their romantic movies, music, dance, fashion, video games, science fantasy, cartoons, and housewife entertainment programs. The export revenue of this cultural industry is enormous for the US government. As of January 2024, it was reported that more than 4 hundred thousand people were working in the motion picture and sound recording industries in the US.

The songs and dances of American singer Michael Jackson are enjoyed by tens of millions of fans across the world. American pop star Taylor Swift's personal fortune exceeds one billion dollars from her singing tours. Her six shows in Singapore from 2 to 9 March 2024 attracted many fans from around Asia. 300,000 tickets were sold at very expensive prices. Her six-night performances are projected to boost Singapore's economy by up to S$500 million (US$371 million) in tourism receipts. This is a huge sum for a population of 6 million people. China's total revenue in the entertainment market is about US$12.14 billion in 2022 compared with US$24 billion in the US. Hence, China with a population of 1.4 billion has a huge potential to catch up with the US.

Many Hong Kong movies are enjoyed by mainland people and overseas Chinese in Southeast Asian countries for many years. Hong Kong kung fu film stars Jackie Chan and Bruce Lee are adored by fans around the world. "Crazy Rich Asians," a 2018 film by Chinese-American director David Chu, became the highest-grossing romantic comedy of the past decade in the US. With a budget of US$30 million, the film grossed US$238.5 million at the box office, making it the first Hollywood production with an all-Asian cast to be the most successful. Most of the film was shot in Malaysia and Singapore. The film received numerous awards, including a nomination for Best Picture in a Musical or Comedy at the 76th Golden Globe Awards and a nomination for Best Actress in a Musical or Comedy for Daniel Wu. It was nominated for the 50th National Association for the Advancement of Colored People (NAACP) Image Award. The film presented the international image of Singapore and Malaysia to the world.

It is a pity that the Chinese film industry had not produced globally famous films until 2025, when 'Ne Zha 2,' an animated movie with a production budget of US$80 million, became an unprecedented smash, racking up $1.9 billion from nearly 80,000 screens after four weeks of release. It's now the highest-grossing movie in a single territory, overtaking 2015's "Star Wars." It is projected that the film will gross more $2 billion receipts soon from around the world.

"Kung Fu Panda" is an American martial arts comedy cartoon series originally released in 2008 as an animated film of the same name produced by DreamWorks Animation. The first two films in the series received Oscar nominations for Best Animated Feature, as well as numerous Annie Awards. The first TV series won 11 Emmys, and the third series won two Emmys. All four films were commercial successes, grossing over $2 billion at the box office, making them the eighth highest-grossing animated film series. This Hollywood martial arts comedy cartoon features a Chinese panda as the main character and Chinese kung fu as the story theme. Hollywood can make successful animated movies with Chinese animals and kung fu. How come China cannot likewise make similar films?

"Crouching Tiger, Hidden Dragon" is a martial arts action adventure film directed by Ang Lee (Taiwan) in 2000. It was adapted from the Chinese novel of the same name. The film consisted of well-known movie stars like Chow Yun-fat (Hong Kong), Michelle Yeoh (Malaysia), Zhang Ziyi (China) and Chang Chen (Taiwan). The film was made in Beijing, with location shooting in Urumchi, Taklamakan Plateau, Gobi Desert, Shanghai and Anji of China. With a production budget of US$17 million, the film was a stunning international success, grossing $213.5 million worldwide. It grossed $128 million in the US, making it the highest box office foreign language film in American history. "Crouching Tiger, Hidden Dragon" won more than 40 awards and was nominated for 10 Oscars in 2001, including Best Picture, and won Best Foreign Language Film, Best Art Direction, Best original score and best cinematography, becoming the most nominated non-English language film at the time. Michelle Yeoh is a movie star born in Ipoh, Malaysia. In 2023, she became the first Asian woman to win the Best Actress Oscar for her multi-faceted performance in "Everything Everywhere All At Once" with many Asian actors.

The two famous Hong Kong movie stars are undoubtedly Bruce Lee and Jacky Chan, who are famous around the world making kung fu movies. Both Bruce Lee and Jackie Chan's martial arts movies produced in Hollywood have promoted Chinese martial arts worldwide for many years. Movie entertainment is an evergreen industry, providing people with many jobs, and is an important Chinese cultural export to the world. Mainland movie producers should cooperate with Hong Kong film and television entertainment companies to produce international movies that attract international movie goers. Many young movie goers around the world like to see Chinese action movies. China should create a movie town in Shenzhen close to Hong Kong to be a China's Hollywood movie town. Good movies could promote China's soft image and Chinese culture. China does not lack good actors and actresses.

3.15 CHINA CREATES 200 NEW SMART CITIES

After lifting 800 million people out of poverty in the past 70 years, China still has 600 million people living in the rural areas. It is an opportune time for the Chinese government to invest 40 trillion yuan to implement a 200 new smart-city project over a period of 10 years during the current period of lackluster economy around the world. The opportunity cost is astronomical if China forgoes this opportunity. China's new smart cities will be considered the most ambitious transformation of the socio-economic political system in human history, creating new smart cities of futuristic design built on artificial intelligence, IoT, robotic autonomous industries and high-tech social infrastructures. Each smart city has an e-government. It is self-sufficient in food and renewable energy. All the 600 million residents will be trained and become the middle income families, paying personal income tax instead of receiving state financial assistance as in the past whilst living in the rural areas.

The budget of 40 trillion yuan is a long term investment with handsome returns and not an expenditure. The government has invested US$1 trillion for the BRI projects for foreign countries in the past 10 years . China has pledged 360 billion yuan (US$51 billion) in fresh funding and promised 1 million jobs for more than 50 African countries who attended the 2024 Forum on China-Africa Cooperation in Beijing in September 2024. This funding will increase China-Africa trade volume and help the Chinese economy. Investment in the new smart city project will likewise generate economic activities.

The smart cities will have a strong knock-on economic effect for the entire country, enlarging the local market for internal circulation economy. The enormous economic and social benefits of a world with 200 smart cities are immeasurable. The government annually spends about $300 billion on crude oil imports, $415 billion on chips, $219 billion on agricultural imports and $13 billion on pork imports. The combined annual expenditure of oil, agricultural products, pork, and chip imports costing $947 billion (more than 6 trillion yuan) will never come back to China to circulate within the economy. This amount of expenditure will increase every year due to inflation and increased consumption.

If this amount of 6 trillion yuan money is invested annually in China, it can create numerous jobs and promote economic development. China's internal circulation economy will generate cascading economic activities for years. The self-sufficient food supply comes from state-of-the-art AI-powered grain farms, livestock and poultry farms, and saltwater and freshwater fish farms. Domestic energy needs come from wind turbines, solar, hydropower, nuclear, biodiesel and tidal waves, to replace fossil fuel and coal as energy production.

All smart cities will have robotic teaching and online medical admission procedures to provide rapid consultation and admission to hospital. The digital yuan will be freely used throughout the country. Using a mobile phone, digital yuan payment does not need internet connection. Unauthorized digital yuan payment will be eliminated as all transactions are traceable. Digital yuan payment makes life convenient and helps eliminate government expenditure on paper money printing and distribution.

China is the first country in the world with a central e-government applying the latest technology and artificial intelligence technology systems to handle provincial administrative affairs in the new smart cities. These administrative affairs encompass economy, society, culture, trade, commerce, animal husbandry, agriculture, banking, finance, education, healthcare, transportation, telecommunications, defense and other fields. With this state-of-the-art system, the administrative data of China's provincial governments are stored on cloud computer servers. The central government has access to the latest information on administrative progress and financial status in each province. AI can track and monitor the progress of various government projects. This will increase the efficiency of China's governance. China will be the largest economy in the world in the near future, with massive application of AI and other technologies.

As China becomes the world's largest economy, state administration will become more complex, and administrative decentralization will become a necessary condition for managing China's vast domestic affairs. Every province will have a long term successive 5-year economic and social development plans. Using the cloud computer and AI, each 5-year plan will go through an annual proof test to validate all the input data for the best economic results to adapt to the prevailing conditions. Every employee only needs to work 25 hours per week, as most industries are operated by robots. All the EV on the roads are autonomously driven. Most restaurants employ robotic cooks.

Xiaomi is building in Changping, a smart factory that operates 24/7 without workers and can produce 60 smartphones per minute or more than 31 million smartphones per year. The quality of production at the factory is controlled by AI controlled machines The smart factory operates without lighting, toilets, canteen, kitchen, air-conditioning or heating, rubbish collection, workers locker rooms, meeting and training rooms, offices and staff car park. This smart factory eliminates problems related to hiring of labor, union friction and grievances, insurance and retirement. This smart factory can cut down a lot of operating costs. Productivity will be the highest in the industry. Such smart factories are yet to appear in the US or elsewhere. China's AI and application in daily life is unparalleled in the world. The US politicians wasting their precious time , energy, and resources to slow down China's economy is an exercise in futility. Chinese people and their government are solidly united and Americans are disillusioned with their government leaders and the American society is fragmented.

3.16 REGIONAL COMPREHENSIVE ECONOMIC PARTNERSHIP (RCEP)

The Regional Comprehensive Economic Partnership (RCEP) was launched by the Association of Southeast Asian Nations (ASEAN) in 2012. It is a free trade agreement amongst 10 ASEAN members and their five free trade partners (China, Japan, South Korea, Australia, New Zealand). The trade agreement was drafted at the ASEAN summit in Bali, Indonesia, in 2011, and formally launched at the ASEAN summit in Cambodia in 2012. The agreement will eliminate about 90 percent of import tariffs between signatory countries within 20 years of entering into force and establish common rules for e-commerce, trade and intellectual property. As of 2020, the 15 member countries account for about 30% of the world's population (2.2 billion people) and 30% of global GDP ($29.7 trillion), making it the largest trading bloc in history. The agreement was signed at the Virtual ASEAN Summit hosted by Vietnam on November 15, 2020, and entered into force in January 2022.

The RCEP is the first free trade agreement signed by China, Japan and South Korea, the three largest economies in Asia. Analysts predict that the deal will help pull the economic center of gravity back to Asia and accelerate the decline of the US in economic and political influence. In 2023, the trade volume between ASEAN and China reached US$ 911.7 billion, making ASEAN and China each other's largest trading partners for four consecutive years, compared with a trading volume of US$783.0 billion with the EU and US$ 664.4 billion with the US respectively. ASEAN had a trade deficit of US$140 billion in 2022 with China. In 2023, China exported to the world approximately US$3.38 trillion worth of goods. Trades between China and Africa, and between

China and countries along the BRI routes, are enormous. China's external trade covers the world without fear of decoupling by the US.

3.17 China Relations with BRICS

Collectively known as the BRICS, an acronym for the five countries , namely Brazil, Russia, India, China, and South Africa. These five countries form an important economic bloc. They account for more than 40% of the world's population and more than 20% of global GDP. These countries account for more than a third of global cereal production. The BRICS countries have an important influence in regional affairs. Since 2009, the BRICS countries have held a formal summit every year. The first formal summit of the BRICS was held in Yekaterinburg on 16 June 2009. The summit focused on improving the global economic situation and reforming financial institutions, and discussed the future of the BRICS and further discussed how developing countries can be more involved in global affairs. The BRICS declared the need for a new global reserve currency that would be "diversified, stable, and predictable."

The New Development Bank (NDB), formerly known as the BRICS Development Bank, is a multilateral development bank operated by the BRICS countries. Its main lending focus will be on infrastructure projects, with an authorized lending capacity of up to $34 billion per year. The bank's initial capital was $50 billion and would increase to $100 billion over time. Member countries will initially contribute $10 billion each, bringing the total to $50 billion. So far, there are 53 projects in the works, worth about $15 billion. The Central Bank of Russia (CBR) has also opened consultations with BRICS countries on an alternative payment system to the US SWIFT system.

China has also begun the development of the Cross-border Interbank Payment System (CIPS), an alternative payment system to SWIFT. It will provide a network that will enable global financial institutions to send and receive financial transaction information in a secure, standardized and reliable environment. The US has used SWIFT as a political tool to sanction countries unfriendly to the US. Since 2012, the BRICS group has been planning to build a system of fiber-optic undersea communication cables for communication between BRICS countries, known as the "BRICS Cable". In August 2019, the Ministers of Communication of BRICS signed a letter of intent on cooperation in the field of information and communication technology. The agreement was signed at the fifth meeting of Ministers of Communications in Brazil on August 14, 2019.

The theme of the 11th BRICS Summit held on 13th —14th November 2019 in Brasília was BRICS "Innovating for Future Economic Growth." The meeting focused on strengthening cooperation in science, technology and innovation, strengthening cooperation in the digital economy, and strengthening cooperation in the fight against transnational crime, especially organized crime, money laundering and drug trafficking. In 2020, Russia, the new interim chair, called for BRICS countries to strengthen their economies, cooperation in the energy and environmental sectors, immigration and peacekeeping resolutions.

As of March 2023, Algeria, Argentina, Bahrain, Bangladesh, Belarus, Egypt, Indonesia, Iran, Nigeria, Pakistan, Saudi Arabia, Sudan, Syria, Turkey, Singapore, UAE, Venezuela, and Zimbabwe have expressed interest in joining BRICS. At the 15th BRICS Summit hosted by South Africa in August 2023, Argentina, Egypt, Ethiopia, Iran, Saudi Arabia and the United Arab Emirates were invited to become full members from January 1, 2024. However, Argentina's new government finally gave up on joining the BRICS. From 2024, BRICS countries will start trading in their own currencies, thereby signaling the start of the de-dollarization process. China has signed currency swap agreements with many countries to achieve bilateral trade without the use of the US dollar and SWIFT system. This will accelerate the use of the Yuan as a currency for world trade. As a result, US financial institutions will lose significant revenue from foreign exchange through the SWIFT system.

The BRICS has introduced a new system in which the members buy and sell directly without using the US system of SWIFT and commodity exchange in Chicago or traders in US or UK. The American farmers will not know the commodity price in the world anymore or the world demand for their agriculture products. The American miners and farmers will be in the dark about their commodity market. The American farmers will suffer the most as to how many crops they need to produce and at what price to make a profit. The US forecast a US$32 billion agriculture trade deficit for 2024. China cancelled many orders of soybean, corn, and wheat from the US and the US farmers do not know which countries have replaced them. The BRICS commodity trade figures are not published publicly.

The BRICS member's direct commodity buying and selling system has crippled the commodity exchange operation in Chicago and London. The financial institutes in US and UK dealing with international commodity payment are bypassed by the BRICS system. Hence, their annual financial losses are enormous. As the BRICS increases its membership to many countries, the de-dollarization will cripple the US dollar as the world trading currency. China cross-border trade will expand significantly when more countries join the BRICS, which is going to be a formidable power group that can challenge the hegemonic USA. Donald Trump said he was not happy with BRICS de-dollarization and threatened a 100% tariff on BRICS.

3.18 SHANGHAI COOPERATION ORGANIZATION (SCO)

The Shanghai Cooperation Organization (SCO) was established on April 26, 1996, in Shanghai by China, Kazakhstan, Kyrgyzstan, Russia, and Tajikistan after signing the Treaty on deepening military mutual trust in border areas. In 2001, the five member states admitted Uzbekistan to SCO. The Shanghai Treaty was proclaimed by six countries on 15 June 2001 in Shanghai and entered into force on 19 September 2003. On 9 June 2017, India and Pakistan formally joined the SCO at the summit in Astana, Kazakhstan. In 2022, Iran was admitted as a full member, bringing SCO to 9 members. The SCO has launched many large-scale projects in the fields of transport, energy and telecommunications, and regularly holds meetings in the fields of security, military, diplomacy, economy and national defense.

Due to its increasingly central position in the Asia-Pacific region, the SCO is widely regarded as the "Eastern Alliance" and the main pillar of security in the region. India's membership in SCO is rather odd. Its foreign policy is more inclined towards the West. It treats Pakistan and China as unfriendly countries over border disputes and involves military confrontation with both countries. On May 7, 2025, Pakistan fighter jet J-10C made in China shot down 3 Indian French made Rafale in a matter of 24- hour dogfight using Chinese missiles PL-15E in an aerial fight over Kashmir. India is also a member of BRICS. India has many trade disputes with China, as India levies heavy fines on Chinese companies operating in India. India joins the US military alliance Quad with Australia and Japan to encircle China militarily in the South China Sea.

In 2021, the SCO member states will account for about one third of the world's population, one quarter of the world's GDP and 80% of the land area of Eurasia. The SCO observer states include Afghanistan, Mongolia, and Belarus. Dialogue partners include Armenia, Azerbaijan, Cambodia, Nepal, Sri Lanka, Turkey; Guest attendance includes ASEAN, CIS, Turkmenistan and the United Nations. In June 2010, the SCO approved the admission procedure for new members. A number of nations also participated in the conference as observers, some of whom expressed interest in becoming full members in the future. In 2012, Azerbaijan, Bangladesh, and Timor-Leste applied for observer status. Egypt and Syria have also submitted applications for observer status, while Israel, Maldives, Ukraine, Iraq, and Saudi Arabia have also applied for dialogue partner status. Bahrain and Qatar have also formally applied to join the SCO.

The SCO Council of Heads of State is the highest decision-making body of the Organization and meets once a year to take decisions and give guidance on all important matters of the Military exercises are also held regularly among the member states to promote cooperation and coordination in the fight against terrorism and other external threats and to maintain regional peace and stability. The Council of Foreign Ministers also holds regular meetings to discuss the current international situation and cooperation between the SCO and other international organizations. The Council of National Coordinators coordinates multilateral cooperation among member states within the framework of the SCO Charter. The Secretariat is the principal executive organ of the Organization. It is responsible for implementing the decisions and decrees of the Organization, drafting proposed documents such as declarations and agendas, serving as a repository for the Organization's documents, organizing specific activities within the framework of the Organization, and promoting and disseminating the Organization's message.

In terms of military activities, the organization's activities have expanded to include enhanced military cooperation, intelligence sharing and counter-terrorism. A number of joint military exercises have been held among the member states to promote cooperation and coordination on external threats such as counter-terrorism and maintain regional peace and stability. India joining this military exercise has a conflict of interest because India is a member of Quad military alliance with the US that is hostile to China. As for economic cooperation, the Vice President of Iran once put forward an initiative that the SCO is a good place to design a new banking system independent of the international banking system. The US applied for observer status in the SCO but was rejected in 2005.

3.19 CHINA FAMOUS PRODUCTS IN THE GLOBAL MARKET

The followings are the well-known Chinese products in the world:market:

- Huawei Mobile Smartphone

Despite the four-year sanctions imposed by the US on Huawei's 5G equipment and 7 nanometer chip exports, Huawei is still able to use domestic production of 7 nanometer chips to produce Huawei's 5G Mate 60 Pro phone. The US government tried to completely strip the phone to determine if any parts were made using products or devices banned in the US, but failed to find any evidence. The US is puzzled as to how Huawei can produce 7 nanometer chips. The Huawei Mate 60 Pro phone has many new features compared to the Apple iPhone. The most startling feature of Huawei's phone is its ability to communicate using satellites when there is no Internet signal anywhere. Therefore, Huawei phones are useful for anyone to contact anyone in the high seas, jungles, earthquake areas, war zones, etc. The CIA can use iPhone to spy on phone users, but not Huawei phones, as it is banned in the US.

- China Electric Vehicle (EV)

In 2024, China's electric vehicle exports exceeded Japan's Toyota for the first time. Such an event has startled the auto industry in Japan and Germany, which are leaders in exporting premium cars. For the first time, Chinese-made EV are finding buyers in Japan, especially since China is not known for making high-quality cars. At the 2023 Munich Motor Show in Germany, there were more Chinese-made EVs on display than any other carmaker model, and Chinese EV were also cheaper than their German-made counterparts. China's EV exports reached 1.3 million units in 2023. BYD is the largest EV manufacturer in China.

In 2023 BYD produced 3.04 million vehicles and were exported globally including Norway, Denmark, Sweden, the Netherlands, Belgium, Germany, Japan, Australia, Singapore, and Malaysia. Its footprint has spread across more than 70 countries and regions, and more than 400 cities. In 2023, BYD reported a revenue of around 483.4 billion Yuan. As of 2024, BYD employs 704,000 people, of which 102,000 are R&D employees. BYD has delivered 70,000 electric buses worldwide, including more than 1,800 electric buses operating in more than 100 major cities in Europe. BYD has a huge presence in more than 30 nations and has a near-monopoly on the electric bus market in the US. There are nine large EV manufacturers in China: BYD, SAIC, NIO, GAC, Li Auto, Geely, Changan, GWM, and Chery. The number of EV registered in China will reach 8.1 million in 2023. BYD overtakes Tesla in EV sales in 2024. In 2023, China controlled 60% of the global market for EV.

- Autonomous Electric Mining Truck

China has started to operate the world's first large scale autonomous electric mining trucks in Inner Mongolia's Yimin open-pit mine. The coal transportation operation is performed by a group of 100 driverless trucks working 24/7 without any human worker and can autonomously load and unload coal. The autonomous electric truck, developed by Huaneng Group in conjunction with Huawei and partners featuring 5G-enabled remote control and autonomous driving, can save thousands of tons of carbon dioxide emissions. These emissions would otherwise be released into the air by conventional diesel trucks. As for charging such a massive power pack, the problem is solved by battery swapping on-site autonomously with a fully charged pack. The empty battery is charged by the on-site photovoltaic green energy source.

Each truck's comprehensive transport efficiency could reach 120 percent of manual operations, and can operate continuously in freezing temperatures as low as -40° C and meet demanding conditions such as high vibration and impact during operation. The autonomous electric mining truck lowers fuel and maintenance costs, and 24/7 operation boosts productivity. These autonomous electric mining trucks eliminate risks to human drivers in hazardous environments, reduce carbon emissions and noise pollution.

- China Generative AI – DeepSeek

DeepSeek, founded in July 2023 by Liang Wenfeng, based in Hangzhou, Zhejiang, focused on artificial intelligence (AI) and big data technologies. The company specializes in leveraging deep learning and big data to develop solutions for industries such as finance, healthcare, e-commerce, and enterprise services. It provides responses comparable to other contemporary large language models, such as OpenAI's GPT-4. The company claims that it trained its V3 model for US$6 million— far less than the US$100 million cost for OpenAI's GPT-4. On 20 January 2025, DeepSeek launched the DeepSeek chatbot—based on the DeepSeek-R1 model—free for iOS and Android. By 27 January, DeepSeek surpassed ChatGPT as the most downloaded freeware app on the iOS App Store in the United States, triggering an 18% drop in Nvidia's share price.

- China High Speed Rail

China has 140,000 kilometers of railway, of which the length of high-speed railway is 45,000 kilometers, the longest in the world. China has established the following railway projects abroad: In 2012,

Addis Ababa-Djibouti Railway standard gauge railway construction was built per the Belt and Road Initiative project. It is an important freight railway connecting the two countries on the African continent, and it is also the first standard gauge electrified railway in East Africa. In September 2018, Mecca—Medina, Saudi Arabia, China built the world's first desert high-speed rail with a total length of 450 kilometers. It cuts travel time between the two cities from 4 hours to 2 hours. In July 2020, construction began on a 1,035-kilometer railway in Laos, a Belt and Road initiative project, as the first standard gauge railway in Laos. It connects the southwestern Chinese city of Kunming with the Lao capital of Vientiane.

The high-speed railway between Hungary and Serbia was opened in March 2022. The Hungary-Serbia Railway is a flagship project of the Belt and Road Initiative jointly built by China and Central and Eastern European countries, with a total length of 341.7 kilometers. The project has cut the travel time between Budapest and Belgrade, the capitals of Hungary and Serbia, from eight hours to three, and marks China's first export to Europe of high-speed trains with speeds of more than 200 kilometers per hour.

In October 2023, the Jakarta-Bandung high-speed railway connecting Indonesia's capital Jakarta and the famous tourist city of Bandung opened, with a total length of 142 kilometers. The journey from Jakarta to Bandung, which used to take more than three hours, now takes just over 40 minutes. It is the first high-speed railway in Southeast Asia and is a project under China's Belt and Road infrastructure initiative

In 2014, China won a contract to supply more than 280 subway cars to the Boston transit system in the US. In 2016, CRRC Sifang won the contract to supply 400 rail cars and has an option for another

446 to replace the largest order of subway cars in Chicago's history. The Chinese company established a factory in Chicago in 2017 and began production in 2019. In 2017, China's CRRC Corp. won contracts to supply subway cars to Philadelphia and Los Angeles. In the Los Angeles deal, CRRC will provide 64 cars for the red and purple lines of the city's subway system for a total price of about $128 million. In the Philadelphia deal, CRRC will deliver 45 subway cars to the US East Coast city for a total price of $137.5 million.

- China Ship Building Industry

China has overtaken Japan and South Korea to lead the global shipbuilding industry, boasting the highest volume of ship construction worldwide. Chinese shipyards are known for producing a wide range of vessels including LNG carriers, cruise liners, nuclear-powered ships, container ships, ore/bulk/oil carriers, and sophisticated naval vessels. The country's shipyards have delivered new models of ships, including the world's largest 93,000 cubic meter ultra-large LNG carrier and a 99,000 cubic meter ultra-large ethane carrier . The Chinese government has heavily invested in upgrading technology and shipyard facilities, aiming to dominate not just in quantity but also in the quality of ships produced. This strategic focus has positioned China as a go-to destination for cost-effective and technologically advanced shipbuilding solutions.

China's better shipbuilding technologies and experiences in construction, coupled with process optimization and digital tools, have shortened construction cycles and improved quality, boosting competitiveness. China's innovation-driven development strategy, focusing on tR&D investment and advanced technologies, has continuously enhanced its independent innovation capabilities, driving product upgrades and technological progress. In October 2024,

Canadian container shipping company Seaspan signed contracts worth around 39 billion Yuan for six 13,600 teu container ships with Chinese Hudong-Zhonghua Shipbuilding company. The contracts have been done in Chinese yuan rather than in US dollars.

According to a report in 2024, China already accounts for 40% of global industrial production and 90% of global shipbuilding orders. The Pentagon's 2023 China Military Power Report said China's navy had about 370 warships. The fleet is expected to grow to 395 ships by 2025 and 435 ships by 2030. The current US fleet is smaller, with about 280 vessels. It expects to reach only 300 in the early 2030s. When the US navy ship needs repair, the turnaround time can vary from a few months to more than a year. According to Govini, the US newest Ford-class aircraft carriers depend on over 6,500 Chinese-sourced semiconductors to operate. Many other US Navy ships and aircraft are similarly dependent on thousands of Chinese semiconductors to function as instruments for US military hardware.

Commercial shipbuilding in the US is virtually nonexistent. In 2022, the US built just five oceangoing commercial ships, compared to China's 1,794. The US Navy estimates that China's shipbuilding capacity is 232 times that of the US. It costs roughly twice as much to build a ship in the US as it does elsewhere. China International Marine Containers Co., Ltd. (CIMC) has been the biggest container-manufacturing company in the world since 1996. It became the first manufacturer with sales volume breaking 2 million TEUs in the global container industry. CIMC's manufacturing operations span 17 different countries, and the company employs over 51,000 people across the world. China's first domestically built large cruise ship, Adora Magic City, weighing 135.5K DWT with a top speed of 22.6 knots and manufactured by Shanghai Waigaoqiao Shipbuilding, completed 84 journeys and carried 600,000 inbound/outbound tourists in its first year of commercial operation in 2024.

- Drone Manufacturing

Da-Jiang Innovations Science and Technology Co., Ltd. (DJI) controls 70% of the global consumer drone market. DJI drones are used in the fields of agriculture, construction and infrastructure, search and rescue, delivery, media, and marketing. DJI reports an annual revenue of over $3.8 billion and is rising. On June 14, 2024 , the US House of Representatives passed a bill that would completely ban DJI's drones from being sold in the US. Most recently, DJI's drones have been used by both sides in the Ukraine-Russia conflict for reconnaissance and bombing. Some American companies tried to replace DJI drones, but other drones were more expensive and their performance unsatisfactory. Even the Pentagon uses DJI drones for their security operations.

In the ongoing Russia-Ukraine war, Chinese drones, originally intended for civilian use, have become integral to military operations in the conflict. Russian media is buzzing with reports of Ukraine using Chinese drones to attack Russian targets. Russian outlets claim they've uncovered how Ukraine can continue procurement of Chinese MAECC drones, sidestepping restrictions from DJI and China's export ban. According to RIA Novosti Russian news agency, Ukraine is receiving these drones through Polish companies and private entrepreneurs, despite official bans. Ukraine reportedly remains an active buyer of DJI MAECC drones, including the MAECC 3 Pro, in defiance of tight export controls and protests from DJI. Ukrainian officials have found American made drones fragile and unable to overcome Russian jamming and GPS blackout technology. At times, the US made drones couldn't take off or complete missions. American drones sometimes fail to fly the distances as claimed or carry the payloads as advertised. In addition, American drones are more expensive than DJI drones. This widespread use of drones has transformed the Russia/Ukraine military conflict into history's first full-scale drone war.

CHAPTER 4: CHINA'S FOREIGN POLICY

A fter the founding of the People's Republic of China (PRC) in 1949, Mao Zedong emphasized that the government's main objectives of foreign diplomacy were centered on national independence, world peace, sovereignty and territorial integrity, and striving for an international environment conducive to China's development. After World War II, the US refused to recognize PRC and even imposed political containment, economic blockade and military threats against PRC. In the face of this situation, China sought friendship with the Soviet Union and other socialist countries to resist US hegemonic dominance. China foreign policy was to attempt to convince all the ex-colonies in the world to form a united front for security and national protection against foreign aggression. In April 1955, the first large-scale Afro–Asian Conference after WW II was held in Bandung chaired by President Sukarno, the first president of Indonesia which was under Dutch colonial rule from 1816 to 1941. It was a meeting of 29 newly independent Asian and African states. The conference's stated aims were to promote Afro-Asian economic and cultural cooperation and to oppose colonialism by any nation. As a young country, China participated in the conference in order to establish closer relationship with the participant countries

4.1 CHINA-US RELATIONS

Good diplomatic relations between China and the US could be traced back to the American Flying Tigers, which helped China resist Japanese aggression during World War II (1939-1945) . Some Flying Tigers pilots sacrificed their lives during operations against Japanese jet fighters. However, the good relationship did not continue when PRC was established. The Kuomintang government retreated to Taiwan in 1949 after being defeated by the Communist Party army. The US government continued to protect Taiwan and prevent the PRC government from occupying the island. The United Nations was established in 1945 with China, the US, Britain, France, and Russia as five permanent members. From 1945 to October 25, 1971, the US sponsored the Kuomintang government to represent China at the United Nations. The US treated PRC as an enemy purely on account of PRC being a "communist" country. The US has a deep disdain against Communism fearing Communism might challenge US political doctrine of democracy.

The first military encounter with the US on foreign policy was when PRC entered the Korean War in 1950 to help North Korea resist the UN forces led by the US. The war lasted until July 27, 1953, when the US failed to achieve its goal in the Korean War of defeating the North Korea army. After the outbreak of the Korean War, the US imposed a trade embargo on China from 1950 to 1972. In 1972, President Nixon visited Beijing, and at the end of his visit, the US and the PRC issued the Shanghai Communiqué, setting out their respective foreign policy perspectives. In the communiqué, the two countries pledged to work towards the full normalization of diplomatic relations. The US trade embargo on China was lifted.

The purpose of American friendship with China was to fight with China against the Soviet Union. President Jimmy Carter announced on January 1, 1979, that the two countries officially established diplomatic relations. In December 2001, the US supported China's entry into the World Trade Organization. Since then, trade between the two countries has grown rapidly, helping China's economy overtake Japan's to become the world's second-largest economy in 2012. Good diplomatic relations between China and the US had reached a climax. In the international diplomatic arena, China remained neutral on international conflicts involving the US.

The US-China relationship had never been more antagonistic since President Trump started a trade war with China, and the relations between the two countries took a plunge even though he kept saying he knew President Xi Jinping very well with good relationships. He even directed Canada to arrest Huawei CFO for violation of US sanction on Iran. He also raised high tariffs on Chinese imports and even destroyed the global supply chain, causing disruption to world trade. Both President Trump and Biden sanctioned export of lithography equipment to China from the Netherlands and banned chip export to China from the US, Japan and South Korea . Although these US sanctions affected Chinese trade with many countries, China's relations with other countries had strengthened through BRICS,SCO and BRI.

In September 2024, Joe Biden approved a US$1.6 billion fund to be used for anti-China propaganda over the next five years. Earlier on, it had moved to close down all Hong Kong Economic and Trade Offices in the US, and issued two advisory notices, one to dissuade American tourists from visiting Hong Kong and the other to dissuade businesses setting up in Hong Kong. The hostility towards China by the US is growing intensively by the day. Meanwhile, the US even considers setting up a NATO-2 in Asia to threaten China, similar to its NATO in Europe threatening Russia. The US forgets that many Asian countries have been the colonies of the US and Europe in the past. It will be interesting to see if Asian countries would want to be their colonies once again to help cripple the Chinese economy when ASEAN and China are each other's important trading partner. The US remains steadfast in its desire to cripple the Chinese economy by hook or by crook indefinitely. The US also remains as an incorrigible troublemaker in the world political arena creating heinous wars around the globe with their 750 military bases, and incessant oppressive sanctions and military threats on other countries including China. In April 2025, Donald Trump imposed a 145% tariff on China imports. On 12 May 2025, China and the US had a meeting in Switzerland and both parties announced a reduction of tariff to 30% by the US and 10% by China. The rollback of tariff reduction would be for 90 days. Often Donald Trump makes decisions in a spur of the moment using his exclusive executive order power.

4.2 CHINA – NORTH KOREA RELATIONS

A t the end of WW II in 1945, Korea was divided by the Soviet Union and the US into two occupation zones along the 38th parallel. Due to political disagreements and influence from their backers, the zones formed their own governments in 1948. The DPRK was led by Kim Il Sung in Pyongyang, and the ROK by Syngman Rhee in Seoul; both claimed to be the sole legitimate government of all of Korea and engaged in limited battles. On 25 June 1950, the Korean People's Army (KPA), equipped and trained by the Soviets, launched an invasion of the south. In the absence of the Soviet Union diplomat, the UN Security Council led by the US denounced the attack and recommended countries to repel the invasion. UN forces comprised 21 countries, with the US providing around 90% of military personnel, entered the Korean War. The US was and still is an intense anti-communism ideology country.

In September 1950, UN forces landed at Inchon, cutting off KPA troops and supply lines. They invaded North Korea in October 1950 and advanced towards the Yalu River that borders China. On 19 October 1950, the Chinese People's Volunteer Army (PVA) crossed the Yalu and entered the war. UN forces retreated from North Korea in December 1950, following the PVA's first and second offensive. PVA captured Seoul in January 1951 before losing it two months later. Following the abortive Chinese spring offensive, they were pushed back to the 38th parallel, and the final two years of the fighting turned into a war of attrition.

The war ended on 27 July 1953 when the Korean Armistice Agreement was signed, allowing the exchange of prisoners and creating the Korean

Demilitarized Zone (DMZ). No peace treaty was ever signed to date. The two Koreas remain as is till to date. Mao Zedong told Kissinger on his 1971 visit to China that he had reason to feel that if he did not stop the US in North Korea, he might have to fight the US on Chinese soil. During the Korean War, President Truman publicly acknowledged the consideration of using atomic weapons against China.

The US dropped atomic bombs on the Japanese cities of Hiroshima on August 6, 1945, and Nagasaki on August 9, 1945, leading towards Japanese surrender to the US and in Asia. China only possessed atomic bombs in 1964 as a deterrence against the US nuclear threat of invasion, as publicly announced. US military threat against China remains till date with 28,500 and 53,700 US troops stationed on South Korea and Japan soil respectively. The US periodically carries out naval exercises in the East China Sea with its military allies like Japan, Australia, UK, and France on the pretext of maintaining freedom of friendly navigation by military ships in international waters. When China possesses nuclear weapons, the US announces to the world that China is aggressively posing a threat against South Korea and Japan's security. However, the US nuclear weapons stationed in Korea and Japan are there only for keeping world peace to protect the security of other nations. So said the US propaganda. Such propaganda shows the US is magnanimous to protect world peace on behalf of other countries in a philanthropic act.

4.3 China — Soviet Union/Russia Relations

W hen the PRC was founded in 1949, the Soviet Union was the first to recognize it. The first recognition from the Soviet Union played a key role in China's world political status, and also provided a good development environment for China's external diplomatic environment. At the end of 1949, Mao Zedong decided to have an official visit to the Soviet Union. Both Stalin and Mao signed a treaty of mutual assistance and friendship, signaling strong support from the Soviet Union to allow the PRC to overcome a series of difficulties in national reconstruction. Stalin died in March 1953, Khrushchev became president of the Soviet Union. The Sino-Soviet honeymoon period under Khrushchev's administration was evident by the Soviet Union's assistance to new China. There were a total of 156 Soviet aid projects for China, of which 109 were new projects led by Khrushchev. Moreover, Khrushchev's aid program to China was more comprehensive and extensive than that under Stalin. Khrushchev also gave China great help in nuclear weapons research. When Khrushchev took full control of the Soviet Union, his attitude toward China began to change. Many times Khrushchev publicly made "anti-China remarks" and many times metaphorical satire, all of which made Sino-Soviet relations slowly deteriorating. In July 1960, under Khrushchev's instructions, the Soviet government announced that all Soviet technical experts would be recalled from China before September 1. Soon after, most Soviet aid projects to China were terminated. Sino-Soviet relations began to deteriorate.

The Sino-Soviet border conflict was an undeclared military conflict between the Soviet Union and China in March 1969 when a group of People's Liberation Army troops engaged Soviet border guards on Zhenbao Island in Manchuria, resulting in considerable casualties on both sides. Further clashes occurred in August at Tielieketi in Xinjiang and raised the prospect of an all-out war. The seven-month crisis de-escalated after Soviet Premier Alexei Kosygin met with Chinese Premier Zhou Enlai in September, and a ceasefire was ordered. To counterbalance the Soviet threat, the Chinese government sought a rapprochement with the United States. This resulted in a secret visit to China by Henry Kissinger in 1971, which in turn paved the way for President Richard Nixon's official visit to China in 1972. Sino-Soviet sour relations continued inconclusively for a decade. Serious talks did not occur until 1991, when an agreement was reached shortly before the Soviet Union disintegration. The border issues were conclusively resolved between China and Russia following a treaty signed in 2003 and an additional agreement in 2008. It is an ironic twist of history that the Soviet quarrel with China propelled China to forge friendship with the US, and now US military threat against China has propelled China to forge friendship with Russia. There are no permanent friends or enemies, only permanent self-interest.

Sino-Russia relations became rosy when Ukraine attempted to join NATO. Its action angered Russia and on 24 February 2022, Russia invaded Ukraine. The Russo-Ukrainian War is the largest conflict in Europe since World War II. After the war had begun, the US and Western countries imposed many sanctions on Russia. Russia could not trade with any country using the SWIFT system. The US and the West also froze Russian bank accounts. The US also sanctioned Russian oil export. Russia began to import all the necessary goods and services from China. Many Russians begin to invest in Chinese Yuan. New Chinese EV are flooding the Russian car market. Russia and China bilateral trade increases many folds. A war lasting more than 3 years will result in tremendous human and financial losses on both sides, and there is no winner in a war. The British Empire won the war against Hitler in WW II but lost the battle when it ceded its global power to the US. The expensive surgical operation was a success, but the patient died.

When Donald Trump became US President (2017-2021), he quietly befriended Putin, asking Putin to declare that China was the culprit spreading the coronavirus. Donald Trump wanted China to pay $10 trillion as reparations to America and the world for the death and destruction caused by the COVID-19. If Putin can do his bidding, Donald Trump would rescind US sanctions on Russia. Putin, a seasoned and acute politician, declined Donald Trump's request. He considered Trump as an untrustworthy politician. Putin is aware of American politicians being : "We lie, we cheat, we steal". Trump can tear off any agreement whenever he pleases. During his first term, Donald Trump signed the USMCA trade agreement with Canada but tears off the agreement in his second term when he signed executive orders imposing a 25% tariff on Canadian imports of goods like automobiles.

American erratic and roller coaster foreign policy changes with the change of president. American politicians are good pot stirrers and uncouth loudmouth. Foreign policy in the US is a joke. Donald Trump picked Pete Hegseth as Secretary of Defense, who has no idea which countries belong to ASEAN. He named Japan and South Korea as members. How long he needs to learn the proper job function of the US Secretary of Defense is a question mark. He even invited his wife to attend meetings with foreign defense officials. He invited a journalist to share in a chat on the discussion of US military plans in Yemen. One wonders what his defense policies are when he is unfamiliar with the geography of the countries around the world and the locations of US 750 military bases overseas.

Michael Loong

4.4 China - Vietnam Relations

China starting in 1950 sent political and military advisers, weapons, and supplies to North Vietnam to help them with their war against the French. After the defeat of the French army by Vietminh at Dien Bien Phu in 1954. Vietnam was split along the 27th Parallel. North Vietnam was led by Ho Chi Minh and South Vietnam by Ngo Dinh Diem. From 1958 onwards, the South came under increasing attacks from the communists within South Vietnam itself. They were called the National Liberation Front (NLF). The US was afraid that communism would spread to South Vietnam and then the rest of Asia. It decided to send money, supplies and military advisers to help the South Vietnamese Government.

In 1963, President John F. Kennedy sent 16,000 military 'advisers' to help the South Vietnamese army. On 2nd August 1964, North Vietnamese boats attacked a US Navy destroyer, the USS Maddox, patrolling in the Gulf of Tonkin. The incident gave the USA the excuse it needed to escalate the war. President Lyndon Johnson got permission from Congress to wage war on North Vietnam. The first major contingent of US Marines arrived in 1965. For the next ten years, the US's involvement increased. By 1968 over half a million American troops were in Vietnam. The war ended in 1975 after years of fighting with the defeat of the South Vietnamese Government. There were 58,220 US military fatalities , mainly consisting of young conscripts, numbering 17,671 soldiers. Many American youths protested in the streets against the US war in Vietnam. The US never learned its own bitter lessons that American soldiers are not keen to fight a war in a foreign country and the war has nothing to do with American national

154

security. Losing human lives to fight an ideology in a foreign country is rather meaningless.

4.5 CHINA-UK RELATIONS

Hong Kong had been a colony of the British Empire since 1841. After the First Opium War, its territory was expanded in 1860 with the addition of Kowloon Peninsula and Stonecutters Island, and in 1898, Britain obtained a 99-year lease for the New Territories. Formal negotiations of the Hong Kong handover began in September 1982 with British Prime Minister Margaret Thatcher in Beijing to meet with Chinese Premier Zhao Ziyang and paramount leader Deng Xiaoping. In 1984, the United Kingdom and China signed the Sino-British Joint Declaration as a treaty between the two governments, setting the conditions in which Hong Kong would be transferred to Chinese control and for the governance of the territory after 1 July 1997. China agreed to maintain existing structures of government and economy under a principle of "one country, two systems" for a period of 50 years.

Hong Kong faced severe pro-democracy violent protests during 2019–2020 . Thousands of young demonstrators waving American flags in the streets burned public buildings. The young rioters also tried to paralyze Hong Kong International Airport by attacking foreign travelers. The activist leaders Jimmy Lai, Martin Lee, Janet Pang and Joshua Wong in 2019 went to US Congress to seek US help to topple the Hong Kong government. Joshua Wong even went to Germany to rally support for his democracy demonstration. The Hong Kong government appeared to have lost control of the riots when the police complained that when the rioters were brought to court for prosecution, the foreign appointed judges let them off without conviction. As soon as the central government in Beijing passed the Hong Kong national security law in 2020, the massive demonstrations came to a halt. Different court judges were appointed to trial all those who committed offenses under the new national security law. It was an open secret that the activists and young demonstrators were financially supported by the CIA.

The China-UK bilateral trade in 2023 reached £107.5 billion (US$ 144.18b). UK exports to China amounted to £38.0 billion (US$50.9b). The bilateral trade is in China's favor. China and UK relations have been uneasy for years. UK sanctions Huawei 5G equipment following the US policy on the pretext of national security. London is home to the biggest offshore RMB clearance center in the world. In 2019, the Shanghai-London Stock Connect and the currency swap program are two examples of successful financial exchanges between the two countries. China-UK relations are further improved by cooperation in the new energy industries: developing battery capacity, offshore wind energy, electric vehicles, and other green technologies. China and UK struck a landmark agreement with £56 billion for construction of high-speed rail networks covering 550 km from London to Birmingham with a speed-up to a maximum of 400 km/h. The project is expected to be completed in 2033. The UK in 2023 received more than 152000 Chinese students. If each student spends US$10,000 per year, the total student expenses would reach US$1.54 billion annually.

4.6 CHINA — MIDDLE EAST COUNTRIES RELATIONS

China oil imports mainly come from the Middle East countries, like Saudi Arabia, Iraq, and Oman. Since the Ukraine war started, China also imported a lot of oil from Russia. For many decades, the Middle East countries had been at war amongst themselves. After WW II, many Middle East countries became independent. However, wars broke out amongst themselves. China's influence in the Middle East had been minimal in the midst of all these wars. The US and the West are the aggressive power in the region, mainly because they want to control the oil trade in the Middle East countries. When the war in Afghanistan ended in August 2021 by the US after failing to win against the Taliban for 20 years and with US$2 trillion down the drain, the US and the West began to withdraw their activities from the Middle East.

In 2018, US imposed severe sanctions on Iran, leading to a sharp downturn in Iran's economy, pushing the value of its Rial currency to record lows, 50% price increase of petrol, quadrupling its annual inflation rate, driving away foreign investors, and triggering street protests. The US also froze the US dollar assets belonging to Iran. Iran sought the help of China to stabilize its economy. In March 2021, China and Iran signed a 25-year cooperation agreement that would strengthen the relations between the two countries that covered political, strategic, and economic areas.

China National Petroleum Corporation (CNPC) was granted an $85 million contract to drill 19 wells in the natural gas fields in Southern Iran. An agreement was reached where China would import 270 million tons of natural gas over 30 years from Iran for $70 billion. Another

Chinese company, Sinopec, would pay Iran up to $100 billion over 25 years for oil and gas purchases and for a 51% stake in Yadavaran field. CNPC also signed a $3.6 billion deal to develop offshore gas fields and then signed another $2 billion contract to develop the northern Iranian oil field near Ahvaz.

China is reported to have helped Iran militarily in the following areas: conducting training of high-level officials on advanced systems, providing technical support, supplying specialty steel for missile construction, providing control technology for missile development, and building a missile factory and test range. It is reported that China is responsible for aiding in the development of advanced conventional weapons, including surface-to-air missiles, combat aircraft, radar systems, and fast-attack missile vessels. On March 2023 Saudi Arabia and Iran announced the normalization of ties brokered by China , with a statement reflecting intentions to resume diplomatic relations between them and re-open their embassies and missions, In addition, the two sides also agreed to implementing cooperation in the fields of economy, trade, investment, technology, science, culture, sports, and youth. China was able to facilitate the diplomatic normalization between Iran and Saudi Arabia, which shocked the US and the world.

4.7 CHINA — EU RELATIONS

The European Union (EU) founded on 1 November 1993 is a political and economic union of 27 European countries. The Union has a population of over 449 million. As of 2023, the EU and China are each other's largest trading partner. China accounts for 9% of EU goods exports and 20% of EU goods imports. In 2023, the EU exported goods worth 224 billion euros (US$250 b) to China and imported goods worth 515 billion euros (US$ 575 b) from China, resulting in the EU recording a trade deficit of 291 billion euros (US$ 325 b) with China. The bilateral trade continues to grow despite the US asking the EU to decouple from China. EU exports to the United States were US$527.49 billion during 2022. In 2023 , EU GDP was US$18.3 trillion, which was larger than China's US$ 17.6 trillion. With US high tariffs on EU imports imposed by Donald Trump in 2025, EU-USA trade will plummet. The EU might seek better trade relations with China. EU members have been wary about US support if the EU does not support US foreign policy wholeheartedly. Some EU members feel that the US has a kill-switch to immobilize their F35 fighter jet. Hence, the EU starts to rearm their defense using only European made weapons.

China and EU foreign relations have been on even keel until the US begins to start a trade war with China. Donald Trump imposed tariffs on steel and aluminum across all nations, including the EU and China. The US pressed the EU hard to decouple from China in trade. It had succeeded in getting the EU to ban the use of Huawei 5G equipment on the pretext of national security. The US also put pressure on Dutch company ASML to stop shipments of some of its ultraviolet lithography machines to China. The US holds substantial power over the EU due to the US holding the absolute power in NATO. The US has troops stationed on German soil since the end of WW II.

The US has substantial influence on the top officials of the European Commission. President Ursula von der Leyen is the present head of the European Commission. She has been calling for the EU to derisk from China, echoing the same tune as the US. She fervently asks the EU to raise tariffs on EV imported from China, following the US policy. The EU has been following the US anti-China smear campaign on Chinese human rights, Uyghurs genocide and Taiwan issues. These stale issues have been the same topics for years whenever China and EU hold bilateral meetings. Germany kept silent when the US was involved in blowing up the cheap natural gas supply pipeline Nord Stream connecting Russia to Germany, which then suffered higher inflation due to higher price of gas import from US and other countries.

4.8 CHINA-FRANCE RELATIONS

France has become China's third-largest trading partner in the EU, with a total bilateral trade volume of $78.9 b in 2023. China exports to France was US$42.1b during 2023, and France exports to China was US$36.8 b. In 2018, there were more than 1,700 Airbus commercial jetliners in service in China, and deliveries to China represent nearly a quarter of Airbus' total jetliner production. China continues to buy Airbus planes. Currently, the Chinese C919 plane has a French engine and some components installed. Airbus has production factories in China. Foreign relations between France and China are considered hostile to China on issues related to Taiwan, Dalai Lama, Xinjiang. In general , France tags along the same US foreign policy towards China. France's anti-China rhetoric is less intense than EU Commission President Ursula von der Leyen's, who is very hawkish toward China with her de-risking from China policy. Her name is not on Donald Trump's favorite foreign friends list.

4.9 CHINA –GERMANY RELATIONS

Germany and China relations remain on firm ground for many years. German Chancellor Angela Merkel visited China 12 times during her 16-year tenure, forging a strong friendship with China. The bilateral trade registered 254.5 billion euros (US$281.6 b⊠ between Germany and China in 2023. China exports to Germany were worth US$100.57 b in 2023. Germany started trading with China as early as in 1978. Its first joint venture in China, Shanghai Volkswagen Automotive Co. Ltd., was established in 1984. BASF, a chemical manufacturer, had an investment worth US$11.07b in a plant in Zhanjiang, Guangdong. China was Germany's largest trading partner for goods in 2023. The EU wants to impose stiff tariffs on Chinese EV imports, but Germany opposed the tariff.

4.10 CHINA — OTHER EU COUNTRIES RELATIONS

China and Hungary relations have been on good terms for many years. Hungary joined the Belt and Road Initiative in 2015. In 2022, Chinese battery company CATL invested US $7.5 billion to build a factory in Debrecen, Hungary. Hungary hosts the largest supply center of Huawei outside of China. BYD will invest 501 million euro (US$559 million) to build EV in Hungary. Geely's Volvo EV is to be built in Belgium. In 2023 Hungarian-Chinese trade was over US$10 billion. In 2022 China built a high speed rail for the Serbian section of the railway with a total length of 183 km, and a designed top speed of 200 km/h. The Hungarian section length will be 159 km and has a top speed of 160 km/h. Thus far, 6.83 million trips in total have been made over the past two years in Serbia. The Hungarian section will be ready by 2025. In 2022, China's export to the Netherlands was valued at $94.5b, and the Netherlands exported $14.6b of goods to China. In 2024, China was restricted by Dutch export controls of ASML chip-making equipment under the US pressure.

4.11 China-ASEAN Relations

ASEAN was established on August 8, 1967, with Indonesia, Malaysia, Philippines, Singapore, Thailand, Vietnam, Myanmar, Cambodia, Laos, and Brunei as members, with a total population of 500 million. After more than 50 years of existence, ASEAN has gained significant economic, social and political progress. ASEAN as a group achieves a significant reputation in the Asia-Pacific region and beyond. The group members receive substantial foreign investments from the US, EU, China, South Korea and Japan. ASEAN Plus 3 was established in 1997 and consists of ASEAN, China, Japan, and South Korea. ASEAN was the first region to sign an agreement to participate in the joint agreement of the BRI, and the ASEAN countries, as a group, were founding members of the China-founded Asian Infrastructure Investment Bank.

ASEAN and China enjoy close diplomatic relations and cooperate in many areas such as economy, security, education, culture, technology, agriculture, human resources, industrial development, investment, energy, transportation, public health, tourism, environment and sustainable development. China's State Council Information Office reported that trade between China and ASEAN reached $982.34 billion in 2024, compared with $688.3 billion in total trade between China and the US. Many Chinese factories that used to manufacture low value supply chain products had shifted to ASEAN countries. At present, ASEAN countries lack many basic social infrastructure, such as high-speed rail, highways, bridges, ports which China can help build. China would build EV in Thailand and Malaysia. Relations between ASEAN and China are on a friendly foundation. ASEAN remains friendly with the US but remains vigilant to the US constant propaganda about

China threat and indirect coercion of joining US military alliance against China. ASEAN countries still remember the atrocious days when they were colonized by the US and the West. Pete Hegseth during the IISS Dialogue in Singapore on May 31, 2025, urged Asian countries to increase defense budget to 5% of GDP to deter China threat. This is the stale US common warmongering theme to promote American military weapon sales to other countries since WWII.

4.12 China-South America Countries Relations

China has established strong political relations with many South American countries like Brazil, Argentina, Peru, Colombia, and Chile. Many South American countries have felt the need to improve their political affiliation with China to balance the political pressure from the US. China engages South American countries through various forums such as the China-CELAC (Community of Latin American and Caribbean States) Forum, promoting dialogue on trade, investment, and development. Over 20 South American countries have joined the BRI, facilitating infrastructure projects like ports (e.g., Peru's Chancay Port) and railways (e.g., Argentina's Belgrano Cargas). Brazil is a member of BRICS

China is the top trading partner for Brazil, Chile, and Peru, and the second largest for Argentina and Colombia. Bilateral trade between China and South America countries reached $450 billion in 2022. Key exports to China are raw materials like soybeans (Brazil, Argentina), copper (Chile, Peru), iron ore (Brazil), oil (Venezuela, Ecuador). Agricultural products are beef, poultry, and wine. South American countries import from China manufactured goods, machinery, electronics (e.g., Huawei's telecom infrastructure), EV, solar panels, and renewable energy technology.

China invests heavily in energy (e.g., hydropower in Ecuador), mining (lithium in Argentina), and infrastructure (e.g., roads in Bolivia). China has exported weapons to several South American countries including Argentina, Bolivia, Venezuela, and Peru. Weapons include radar systems, fighter jets, and patrol ships. According to a report, between 2000-2020, Chinese weapon exports to South American countries grew 26-fold from $17 billion to $315 billion. With more shipping lanes, megaports and logistics corridors linking China and Latin America across the Pacific, trade between China and the region reached $518.4 billion in 2024, doubling the figure recorded a decade ago. The population of South America in 2024 is 434 million, compared with the USA population of 340 million.

CHAPTER 5: CHINA GLOBAL TRADE
5.1. CHINA WORLD TRADE INITIATIVES

China has been trading with other countries for over 2000 years. China once had the highest standard of living in the world, with half of the world's gross domestic product. China had been an agricultural country since ancient times, but there were also crafts, ships, silk, ceramics, household products manufacturing, as well as mining, transportation, construction and other industries. Under a monarchical government, the people were free to start their own business. In 139 BC, China first opened the western regions from Xi'an to India and Iran along the Silk Road. Chinese silk, iron smelting, well drilling, paper and other technologies were spread to Europe. It was reported that in the early 17th century, Dutch traders started to import tea, teapots, tea bowls and saucers from China to Europe. The Silk Road enhanced the progress of China's foreign trade and made China the world's largest economy. Towards the end of the Ming Dynasty, the Ming government gradually closed its doors to foreign trade, isolating China from the outside world. China's economy began to decline.

Britain enjoyed rapid economic progress brought about by its First Industrial Revolution with the invention of the steam engine. Although Britain had a small population, its citizens had an active interest in investing in research and development, making ships, aircraft, cars, and industrial and commercial products that were sold around the world. Britain became the world's largest economy. However, after WW II, the British economy collapsed, and Britain's strong economic position was ceded to the US. Many European elites immigrated to the US, and

American industry began to flourish and produced many products sold worldwide. Soon it became the world's largest economy.

China opened its door to the outside world in 1978, welcoming foreign investors to build factories in China. From initially producing low value exports to gradually upgrading to producing higher value-added export products, China finally attained the status of the world's second-largest economy in 2012. The US is not happy that the two-way trade with China is favorable to China each year. The US viewed China's high economic growth as a threat to US security and began to adopt trade protectionism against Chinese imports by raising import tariffs and selective trade sanctions.

5.2 IMPERIAL CHINA'S SILK ROAD TRADE

International trade entrepreneurship originated from China's overland Silk Road. The object of trade was to sell goods to those willing to buy them at a price, and the two parties were satisfied with the exchange of silver currency for goods. In the past, the volume of international trade over the Silk Road was small because the weight of goods carried by the camels over long desert distances was limited. The Silk Road was the earliest and most important communication channel between China and Western civilizations. It originated during the reign of Emperor Wudi of the Western Han Dynasty (202-8 BC), with its capital Chang'an (present Xi 'an) as its center. The Silk Road opened up international trade across the Eurasian continent, promoting international cultural, economic, scientific, industrial, commercial and social exchanges.

In addition to ceramic products and tea, silk was the most valuable Chinese export. Silk was the noble fashion goods of aristocrats and wealthy families in many European countries. Silk was invented by the Chinese. As Britain was short of silver coins to continue buying Chinese silk products, it began to introduce opium for sale in China to earn silver coins. As the Qing imperial court objected to the opium sale, a war erupted between the two countries. In 1839 the Chinese government confiscated and destroyed more than 20,000 chests of opium in Guangdong.

In June 1840, 47 British ships and 4,000 troops arrived at the Pearl River, sealing off the sea and opening the first Opium War. The Qing imperial government was defeated in the war, and in 1842 it ceded

Hong Kong to the British Empire. Later The Eight-Nation Alliance, comprising Russia, France, Japan, British Empire, Germany, United States, Italy and Austria-Hungary, invaded northern China in 1900, destroying Beijing's Forbidden City. China became a semi-colony. In 1911, a group of revolutionaries led a successful revolt against the Qing Dynasty, establishing in its place the Republic of China, thus ending the 2000-year-old imperial system in China. The invention of silk changed the history of China and the world.

Tea has been consumed in China as a beverage as early as 4,000 years ago. During the southern Song dynasty (1127–1279), Arab merchants acquired tea from the city of Quanzhou in Fujian Province and carried it to the Middle East and other countries via the Silk Road. In 1610, a Dutch ship calling at Macau took the first load of Chinese tea to Europe. By the early eighteenth century, Europeans had come to view tea-drinking as a symbol of wealth and sophistication. Tea was exported by the British East India Company to the British Colony in North America. On 16 December 1773, 342 chests of Chinese tea belonging to the British East India Company were thrown from the ships into Boston Harbor by American patriots, an event that has gone down in history as the Boston Tea Party. This political and mercantile protest was one of the key events that led to the American Revolutionary War and finally American independence. Chinese tea changed the history of the British Empire and the world.

The maritime "Silk Road" via the South China Sea extended to the Mediterranean countries of Europe. Admiral Zheng He (1371-1433) of the Ming Dynasty, a eunuch and a Muslim, set sail for the first time in 1407 with more than 270,000 people. During his seven voyages to the seas, he reached more than 30 countries in present-day Southeast Asia, Africa, and South Asia. These records represented the peak of China's maritime exploration, more than 80 years before the Western explorers such as Vasco da Gama Columbus. At that time, the Ming Dynasty was ahead of the European countries in navigation technology, fleet size, distance of voyage and duration. Unfortunately, after Zheng He ended his 7th sea voyage, the imperial court stopped his expedition. Emperor Zhu Yuanzhang implemented a close-door policy to stop the outflow of silver coins. His policy was designed to prohibit trade between Chinese and foreigners, thereby retarding China's future economic progress for more than six hundred years.

5.3 Trade Disputes Between the US and Europe

The US has been engaging in trade wars with many countries before WW II. From 1929 to 1939, it was the longest and deepest recession in US history. The Great Depression was triggered by the stock market crash of 1929. The collapse of world trade was a result of the 1930 Smoot-Hawley Tariff Act creating bad economic policies, bank failures and panics and the collapse of the money supply. The Great Depression in the US affected almost every country in the world and was characterized by sharp declines in industrial production and prices, mass unemployment and banking panics. In 1933, one in five banks failed. Poverty and homelessness rates had risen sharply. During the Great Depression, industrial production fell by nearly 47 %, gross domestic product fell by 30 %, and unemployment reached more than 20 %.

The Smoot-Hawley Tariff Act, which raised tariffs to their highest levels in the 20th century, was intended to protect the US economy by forcing Americans to buy only American-made goods. High tariffs would make selling goods so expensive that foreign countries would have to stop exporting goods to the US. This is a form of economic protectionism. The Act was meant to protect the US economy by not allowing Americans to buy goods from outside the US; but it ended up making the economy worse. In response, the European Community also began to impose high tariffs on American imports. It was the export demand associated with World War II and expanded government spending that restored the US economy to full employment productive capacity in 1941, finally ending the Great Depression.

In the 21st century, the US and the European Union have faced a series of complex and high-profile trade disputes. The two parties constitute the largest and most important bilateral trade relationship in the world. More than $1 billion in goods, services, and investments flow between the two regions every day. The two parties account for about 50% of each other's outward direct investment. Some trade disputes between the US and EU had reached tipping points. The main disputes were: steel, beef, genetically modified agricultural products such as genetically modified corn, soybeans, tomatoes, wheat, bananas, and aircraft manufacturing subsidies. The settlement of these disputes took years of marathon discussions at the WTO. In March 2018, Trump imposed tariffs on 25 percent of steel and 10 percent of aluminum imports from the EU.

In 2004, the US brought a case to the WTO accusing the European Union of illegally subsidizing Airbus, The European Union also filed a lawsuit in May 2005 against the US for illegal subsidy to Boeing. The WTO eventually ruled that both sides had broken the trade rules, and since then both have taken steps to cancel subsidies found to be at fault. The US and the European Union reached a ceasefire agreement in 2021 in a 17-year trade dispute over Boeing and Airbus subsidies. Under the agreement, the two sides will eliminate taxes on $11.5 billion of goods over five years, including wine, cheese, and tractors. The tariffs were imposed as punishment by both sides in an escalating dispute.

5.4 US-Japan Trade War

The trade conflict between Japan and the US dated back to the 1930s, when increased exports of Japanese textiles to the US led to accusations of unfair competition. The Japanese government agreed to limit the amount of certain Japanese goods, mainly cotton textiles, to be exported to the US. Almost all US-Japan trade conflicts arose from complaints about surging imports of Japanese goods into the US market. American complaints about lack of access to the Japanese market had been around for a long time. As a result, it eventually provoked a protectionist reaction in the US, leading to friction between the two countries. The bilateral trade conflict between Japan and the US in the 1980s and early 1990s was marked by a shift from Japanese exports to negotiating greater US access to the Japanese market. The US complaints centered on the following items:

- Textiles ---- Textiles, which had been Japan's main export before World War II, would become Japan's main export again in the early 1940s. The increase in Japanese cotton exports to the US in the mid-1950s caused the American textile industry and labor unions to complain and protest against Japan protectionism to the US government. Instead of unilaterally imposing measures such as import quotas, the US government has asked Japan to set limits on its exports. This was the beginning of a long history of voluntary export restrictions.

- Agricultural products ---- The US was upset that Japan restricted beef imports in order to protect its own products. In the late 1970s, the US demanded that Japan open its beef market. Japan eventually consented and lifted restrictions on US beef imports

- Cars ---- During the 1973 Arab-Israeli War, the Arab members of the OPEC imposed an oil embargo on the US on October 17, 1973, in retaliation to American oil supplies to the Israeli army. At the same time, OPEC began a series of oil production cuts that nearly quadrupled the price of oil, from $2.90 per barrel before the embargo to $11.65 per barrel in January 1974. American car owners began to buy fuel-efficient vehicles to reduce their daily living expenses. American automakers didn't produce fuel-efficient cars, so Japan took the opportunity to export many fuel-efficient cars to the US market. In 1973, only 740,000 Japanese passenger cars were sold in the US, accounting for 6.5 percent of the market, but sales nearly doubled by 1977,reaching 1.9 million units by 1980. By contrast, the US auto industry was in deep decline. In two years, production fell 30%, 300,000 autoworkers were laid off, and half a million jobs were lost in auto parts, steel, and ancillary industries. The auto conflict was sparked by the aggressive entry of Japanese cars into the US market. Under intense pressure from the US, Japanese automakers introduced a Voluntary Export Restriction (VER) in 1981, agreeing to implement a voluntary export restriction of 1.68 million vehicles per year.

- Semiconductor ---- The semiconductor dispute originated from the Japanese government's traditional strategy of promoting and protecting a highly competitive domestic industry. By the mid-1980s, Japan had surpassed the US in both quantity and quality of semiconductor production, and Japanese companies had replaced American companies as the leading commercial semiconductor producers. As Japan's share of the global semiconductor market had risen significantly, American chip manufacturers were at a significant relative disadvantage. As Japan and the US clashed over the semiconductor trade issue, the US took a hard line, with their negotiations among the longest and most intense between the two countries. Finally, under sustained US pressure, backed by threats of further action, led to major bilateral agreements that gave American chipmakers some relief from Japanese dumping in the US market and greater access to the Japanese market. In January 1987, the Office of the US Trade Representative threatened to reimpose trade sanctions if Japanese companies failed to comply with the terms of the agreement by April 1. On March 27, 1987, the US announced a 100% retaliatory tariff on $300 million of Japanese electronic goods. According to a September 2020 report by the Semiconductor Industry Association, in 1990, the US produced 37% of the global supply of chips. But now, the country accounts for just 12%. According to the report, 75% of the world's chip manufacturing comes from Asia, and China will be the largest chip producer by 2030.

5.5 JAPAN-EUROPEAN COMMUNITY (EEC) TRADE FRICTION

In the 1930s, when many countries suffered from the US Great Depression, Japan's exports expanded significantly around the world, leading to serious trade conflicts. In the second half of the 1970s, when Western countries suffered a severe economic depression, the European Community began to ruthlessly criticize the inflow of Japanese goods. At that time, the best-selling imported Japanese tape drives had been adversely reacted in the European Community. Another source of dissatisfaction with Japan was its lack of reciprocal treatment of EC exports. In 1986, Japan exported 1.1 million cars to the European Community, while importing only 66,000 from the EU — a ratio of 17 to 1. The EC was therefore very unhappy with Japan's large trade balance. Finally, Japan agreed to voluntary export restrictions.

5.6 China-US Trade War

In 2010, the Obama administration had been brewing the threat of a trade war with China for some time. The US government , using its traditional method of showing unhappy trade relations with another country like Japan, complained that the Chinese Yuan was undervalued, leading to a large Chinese trade surplus. In November 2017, Trump paid a "state visit" to China. On March 22, 2018, Trump signed a bill limiting American investment in key technology sectors in China, and imposed tariffs on Chinese products related to aerospace, information, and communications technology and machinery. On April 2, 2018, China imposed tariffs on 128 products from the US, including fruit, wine, seamless steel pipe, pork, and recycled aluminum in retaliation for the US tariffs on Chinese steel and aluminum.

On April 16, 2018, the US Department of Commerce concluded that the Chinese telecommunications company ZTE had violated US sanctions. US companies were banned from doing business with ZTE for seven years. From May 3 to 7, 2018, China and the US held trade talks in Beijing, and the US demanded that China reduce its trade deficit by $200 billion within two years. The talks ended without a resolution. On June 15, 2018, the US Customs and Border Protection announced a 25% tariff on 818 Chinese imports worth $34 billion. In July 2018, Trump followed through on sweeping tariffs on China imports for alleged unfair trade practices. US Treasury Secretary Steven Mnuchin and Chinese Vice President Liu He had opened talks on a trade deal. In the months that followed, the two countries became embroiled in countless back-and-forth negotiations, a tit-for-tat tariff war, and several WTO cases that eventually brought US-China trade tensions to the brink of a full-blown trade war. Donald Trump was determined to engage in a full-blown trade war with China.

Finally, on January 15, 2020, the two sides signed the Phase One agreement. The agreement formally agreed with China to reduce tariffs, expand trade purchases, and recommit to intellectual property, technology transfer, and currency practices. The final agreement included US tariffs on Chinese goods totaling $550 billion and Chinese tariffs on US goods totaling $185 billion. China agreed to buy $80 billion worth of US agricultural products in 2020 and 2021. On February 17, 2020, China granted tariff exemptions to 696 products, including pork, beef, soybeans, wheat, corn, sorghum, ethanol, liquefied natural gas, crude oil, steel rails and some medical equipment. There was still room for maneuver in the negotiations between China and the US on the trade war issue.

The grand strategy of the US to contain China's rise would not change, and the trade war against China's high-tech field would only become more intense. In 2024, Joe Biden introduced new trade measures against China by imposing tariffs on EV with 100%, semiconductors 50%, solar panel 50%, battery 25%,steel & aluminum 25%, port crane 25%, rubber & surgical gloves 25%. Such measures also fueled inflation in the US. On 2 April 2025, Donald Trump issued a global crusade of tariffs on every country. He imposed a tariff of 145% on Chinese imports. After many protests from the industries and public, he reduced the Chinese tariff to 30% for a 90-day reprieve.

The two countries have not negotiated a second phase of a trade deal since Trump left office in 2021. What Trump may not have realized is that setting sanction limits on Chinese high technology could accelerate China's march towards the age of robotics, biotechnology and quantum computers. In any case, China has been careful with the US trade war to avoid disentangling the diplomatic relationship between the two countries, which has a great impact on China's economic progress. China's economy must move in the direction of growth. The American people have no hostility towards the Chinese people. American businessmen still like to continue investing in China because their investments are profitable. American industry in China has been stable for many years, and there is no reason to withdraw from the Chinese market and move to other unfamiliar countries.

5.7 Impact of US Tariff Policy on International Trade

The US has demanded China to address the trade imbalance , including by removing some industrial policies that the US considers unfair and harmful. Trump's trade war has also aroused public opinion and political opposition in the US, fearing that this trade war will affect the economic growth of the US. China's American exports contain parts from international supply chains of materials, and some of China's exports are made in China by American companies. Hence, China-made products do not compose completely native Chinese materials. Economic experts have warned that a trade war between the two economic powers will inevitably affect the global trade environment and undermine world economic growth. The world has been interconnected for many years through a global supply chain, which can become chaotic when one part of the chain is broken.

The top five trade deficit countries with the US are China at $375 billion, Mexico at $71 billion, Japan at $69 billion, Germany at $65 billion, and Canada at $18 billion. On a per capita basis, the US trade deficit with each country is $792 with Germany, $547 with Japan, $537 with Mexico, $486 with Canada, and $267 with China, bearing in mind China has a huge population. Germany makes the most money from the US. The US should make the strongest possible protest against the German large trade deficit based on per capita basis.

The US investors will not let Sino-US trade deadlock affect their operation in China, thinking that the US government will never dare to fight this trade war to the end, and eventually becoming a rotten apple. It is obvious to all that today's world trade has entered an era of unprecedented integration, not only the production division of labor becoming more and more detailed, but also almost every country's high-end products assembled with components made in different countries. According to an article in the US National Interest magazine, most observers believe that the trade war with China is impossible to win, and the US will not achieve the desired goal by adopting unilateral tariffs against China. Moreover, doing so could have long-term negative effects, reinforcing China's belief that it will prevail in future bilateral contests.

The current US belligerent foreign policy against China and irrationality of massive tariffs on foreign imports beget early policy failure, similar to the saga of failure of the 1930 Smoot-Hawley Tariff Act, which raised tariffs of 40%to 60% on some European imports. The Trump Administration's misguided and cowboy-culture economic policies would affect global economic progress. The use of the US dollar as a political tool to sanction several countries leads to disruption of international trade, as has happened to Russia arising from the Ukraine war. Russia fertilizer and natural gas exports are curtailed in the world market. China's manufacturing industry is more extensive than that of the US, and China can sustain in maintaining multinational exports to other countries and survive, while higher prices of American exports will hit its economy hard. The high price of imported goods will affect every citizen's pocket, and could trigger inflation, economic recession, and increased unemployment. This situation contradicts Trump's desire to make America great again.

Trump wants all American factories to be relocated to the US, forgetting that all the American factories in China depend on the Chinese supply chain to make their products. The US relies on China for the supplies of rear-earth minerals to manufacture their high-tech weapons. Trump failed to cripple China's economy during his first term of presidency. It is doubtful he can succeed this time as China's present economy is more solid than 8 years ago in 2017 when Trump took office. He is now sitting on a national debt of US$36 trillion and rising, and simultaneously beset by two frustrating money devoured wars in Ukraine and Palestine.

5.8 TRADE WAR SPURS CHINA CHIP INDUSTRY DEVELOPMENT

In 2018, the US started a technology war against China. Technology includes seven fields: 5G, semiconductors, artificial intelligence, supercomputers, genetic technology, military weapons, green energy technology, and the core of trade dispute is the semiconductor chips. The Biden administration established the CHIP 4 alliance in July 2022, including the US, Japan, South Korea and Taiwan. The US arm-twisted Taiwan TSMC to build a factory in Arizona to manufacture 3-nanometer chips, as well as Samsung to build plants in Texas. On August 9, 2022, Joe Biden passed a bill with an allocation of US$ 11 billion fund dedicated to advancing semiconductor R&D. The US dominates the semiconductor design sector with companies like NVIDIA, Intel, and Qualcomm.

On August 31, 2022, Joe Biden issued an injunction requiring Nvidia to prohibit the sale of some high-performance GPU chips to China. AMD is also subject to a related ban. These chips are mainly used in AI, data analysis, medical and scientific computing. As a final step, the US will force the Asian chip factories in the US to sell to American companies, thereby enabling the US to control global manufacturing of high-end chips. This should stifle China's technology advancement.

In 2024, the Chinese government established a national fund worth $29 billion to assist in the semiconductor industry. It urges enterprises to accelerate innovation. In response to the US sanctions, the Chinese government offers tax-free incentives to Chinese semiconductor manufacturers to develop chips that are 7 nanometer and smaller. This should spur China to accelerate the development of advanced technologies, making China self-reliant in the microelectronics industry. In 2023, Huawei was able to use a SMIC-produced 7 nanometer chip to power its new Huawei smartphone Mate 60 Pro that shocked many critics in the US.

5.9 China Bans Export of Rare-Earth Minerals to US

The Joe Biden administration spends billions of dollars to help local companies to build factories to manufacture batteries and semiconductors which need graphites and anodes that are mainly produced in China, and China has imposed strict control on exports of graphite and anode to the US. US EV manufacturers face problems with the supply chain of batteries and graphites. China controls mining and processing of raw minerals into final elements for chip, battery, and semiconductor manufacture. China mines about two-thirds of the world's natural graphite, controls about 60% of synthetic graphite, and almost 100% of coated spherical graphite used in EV batteries. China also accounts for 98% of anode manufacturing capacity expansion up to 2030 according to the International Energy Agency.

Although Joe Biden has banned US semiconductor exports to China, according to Govini report, the US newest Ford-class aircraft carriers depend on over 6,500 China-sourced semiconductors to operate. Many jet fighters are similarly dependent on thousands of Chinese semiconductors to function. China places a ban on exports of gallium and germanium to unfriendly countries. In 2023, China controls over 85 percent of the world's rare earth processing capacity, and approximately 417 kilograms of these are required to produce each F-35. On 7 Oct 2022, F-35 fighter jet deliveries were temporarily halted following the discovery that a subcontractor had made a magnet using a cobalt and samarium alloy that was sourced from China. Flat panel displays are also sourced from China, which are no longer made in the US.

Some analysts questioned in 2018 whether it was even possible for the US to make purely American weapons, highlighting the low feasibility of re-shoring production to the United States. The complete supply chain for US military equipment production is not available in the US. China is the leader in the world supply chain and its industries can produce components for dual civil and military applications and this is unavailable in the US. Yet Donald Trump in 2025 wanted to increase tariffs on Chinese imports including rare earth minerals, forcing China to ban the exports of these minerals to the US. His global crusade of high tariffs is contrary to his MAGA policy.

5.10 CHINA CONCENTRATES ON HIGH VALUE PRODUCT PRODUCTION

The US trade war with China provides China a good opportunity to rise to the top of international leadership. China has spearheaded the SCO, BRICS, and the Belt and Road Initiative, actively expanding its trade scope, increasing its international power and political influence. This leadership aims to foster a new China characterized by numerous private high-tech industries, advanced e-government administration, food self-sufficiency, unlimited water supply, pollution-free transportation, and robotic education and medical systems. A cashless society and the use of renewable energy eliminates oil imports and greenhouse gas emissions. In a few years, Chinese-made battery electric vehicles will become the world's largest production, and the market will cover the world. It is an unprecedented event that China's EV world market overtakes those of Germany and Japan in a span of 10 years.

China needs to maintain friendly diplomatic relations with the rest of the world to promote China's rapid economic reform. Every policy, every product, every system has an expiration date. Old products can be obsolescent by new ones quickly. The old production lines are not as efficient as the new robot-operated ones. Many well-known and long-standing companies in Western countries become obsolete as time goes by as they have not upgraded to improve productivity and efficiency. China bypasses the use of credit cards by adopting digital payment using Alipay and WeChat. Such adoption of digital payment saves time, human resources and improves social development, and also provides convenience to daily goods and services transactions between customers and vendors. The government does not need to print so much paper money for circulation and the banks do not need to count daily cash movements. US sanctions against China are a catalyst for China's scientific and technological innovation and advancement towards a futuristic world.

Now China wants to build itself into the most modern technological and industrial power, turning "Made in China" into "Created in China" high-quality products. China's core projects involve establishing an electronic operating administrative system for the government, leveraging the Internet, IoT, big data, cloud computing, and AI. These initiatives extend to robotic education, robotics, telemedicine, and high-tech development, alongside fostering private R&D to create new businesses. Additionally, China aims to attract foreign elite and talent, and build a robust, advanced electronic remote national defense force.

To cope with this ever-changing world. China needs to train more diplomats to deal with international affairs and financial experts to manage the Chinese Yuan as a global trading currency. China actively refrains from getting involved in political turmoil around the world created by the US. Many Chinese factories are operated by robots and AI. Even the famous Chinese radar guided air-to-air PL-15E supersonic missiles (costs about $200,000 each) that were used by Pakistanis air force on 7 May 2025 to shoot down three Indian French Rafale fighter jets (costs $240m each) are made by cutting-edge robotic production lines. In February 2023 the US air force used Sidewinder missiles (costs about $500,000 each) to shoot down a Chinese weather balloon costing less than $100. In May 2025, a US F35 fighter jet nearly got shot down by a surface-to-air missile fired by the Houthis in Yemen at the Red Sea.

5.11 OUTCOME OF THE SINO-US TRADE WAR

No matter how the US reacts to China's economic development, China continues to improve its economy and technical development. No matter what moves the US makes, China has the determination and patience to continue along its own established course. The trade friction can only last for a period of time, as proven by history. The imposition of tariffs on imports also affects American businesses and has a great impact on the American economy and living standards. International trade relies on global supply chains to produce cheaper goods for American consumers. Today's world trade has entered an era of unprecedented integration, and as the US is a capitalist country, American investors have the freedom to invest in any country that they feel they can make money. For example, the US president cannot order Apple to stop manufacturing iPhone in China, nor tell Elon Musk to stop building EV in China or McDonald's not to sell hamburgers in China. Many American companies have made a lot of money doing business in China in the past. GM sells more cars in China than in the US. China used to spend hundreds of billions annually to buy computer chips and other electronic products from US companies until both Donald Trump and Joe Biden put sanctions on their exports.

If the US continues to impose sanctions on exports to China and raising tariffs on Chinese imports, such a policy will hurt bilateral trade. The US consumers will pay higher prices on imported goods made in China. This bilateral trade is a win-win for both China and the US, providing jobs for Chinese workers and cheap goods for consumers in the US and other countries. China has signed bilateral trade agreements with many countries, and Chinese products can still sell to many countries around the world. The Chinese market is the largest in the world, and hence American companies will not like to lose such a high market.

The US politicians do not realize that many American companies still need to import many components and parts from China to assemble their final products in the US. Even the Pentagon wants to import drones made by Chinese firm BJI, although import of the DJI drones are banned by Joe Biden. The navy shipyards need to import semiconductors from China to repair their navy ships. The latest jet fighters F35 have Chinese materials built into the plane. Even Apple has to import hundreds of thousands of small screws for its iPad production in the US.

The many generic medicines the US bought from India have Chinese chemicals active pharmaceutical ingredients (APIs). The US imports many medicines from China. China accounts for 70% of the US imports of paracetamol. This fact is probably unknown to US politicians, as they do not buy medicines personally. According to data from the US Commerce Department, in 2018, China accounted for 95 percent of US imports of ibuprofen, 91 percent of hydrocortisone, 40 to 45 percent of penicillin, and 40 percent of heparin. China is a major supplier of APIs for many prescription drugs and antibiotics consumed in the US. It is not easy for the US to decouple from China on medicine supply. It is inconceivable that the US decoupling from China can succeed without serious consequences similar to the 1930 Smoot-Hawley Tariff Act that worsened the US great depression.

When Donald Trump introduced a 145% tariff on Chinese imports on his liberation day of 2 April 2025, many US companies stopped ordering goods from China. When Walmart and other shopping malls complain about their shelves getting empty soon and prices of remaining stocks are rising, many consumers are worried. When thousands of docker workers are laid off on the docks because no container ships are arriving from China, the dire situation sends a strong signal to Donald Trump to get his tariff removed. Suddenly, he directed Secretary of Treasury Bessent to seek an urgent meeting with a Chinese official to begin a dialogue. After two days of meeting in Switzerland, and on 12 May 2025, Bessent announced the US would lower the tariff from 145% to 30% rolling back for 90 days. Further discussions will follow in future on trade issues before the expiry of 90-day reprieve.

China's industries are quite mature. Private and state-owned enterprises have the ability to create new industries and products on their own without the need for foreign cooperation. China's years of low-margin industries are overdue, and it is time to produce high-quality, high-priced branded goods, turning "Made in China" into "Created in China" logo. The central government has long-term continuous 5-year plans for implementations of new industrial and economic renewal measures, not only for self sufficiencies in foods, energy, and semiconductor chips but also for exports. The EVs made in China have now swamped the world for the high-tech design and features at low price and high quality.

The scope of the EV can be further enhanced to increase new features for additional customers needs, like an autonomous EV taxi that can accommodate a handicapped person to be sent to a hospital A&E ward. Perhaps China can produce autonomous electric river boats for transportation in the river. China has begun to produce robotic chefs to cook Chinese foods. Soon, China can invent a robot that can cook western cuisine. China will produce new schools that can teach students by robotic teachers and IT classrooms with real time response to the students' questions. Generative DeepSeek AI will enhance education standards in Chinese schools.

The US trade war with China has nothing for the US to gain except the successive US governments continuing to waste precious time trying to cripple China's economy. It is an exercise in futility. The US politicians must realize that China is a huge country with 1.4 billion people possessing 5000 years of history with a few moments of glories of being the largest GDP in the world. China is not Yemen or Afghanistan or Iraq. China can design and manufacture a space station that the US is yet able to produce. Many Americans do not realize there are more billionaires (814) in China than in the US (800) in 2024, and China PPP GDP is 26% higher than the USA in 2023. It is a pity that the US politicians are attempting to cause the country to fall into oblivion.

Chapter 6 The World Entering a Digital Age
6.1 China Digital Economy

The world today is undergoing the most rapid, extensive and profound changes in human history. High technologies, represented by IT, AI, IoT and big data, are advancing by leaps and bounds. The competition in national strength, which is mainly characterized by the information industry, is becoming increasingly aggressive. The profound impact of informatization on economic development and social progress has aroused widespread interests around the world. Both developed and developing countries attach great importance to IT application and take accelerating IT application as a strategic task for economic and social development. The best example of utilization of digital transformation is the latest EV built in China that shocks the European and Japanese legacy carmakers. Chinese EV are equipped with loads of IT software, batteries, sensors, computer chips, servo motors, digital information and IoT technology. Chinese EV outsells Tesla in China.

China has more robots installed in factories than any other country. The new Xiaomi Pengpai AI-powered factory serves as the company's AI brain, enabling the production system to autonomously develop, optimize processes, and address production issues in real-time. This allows the factory to operate 24/7 without human workers, producing 60 smartphones per minute. This AI-powered autonomous factory is the first in the world with the highest productivity and quality. The IT created by the digital revolution is a strategic industry. It can carry out both manufacturing activities and service businesses, or engage in both simultaneously, thus becoming a mixture of manufacturing and service industries. The average annual growth rate of the world economy was about 3%, but the growth rate of IT and related industries was 2–3 times that of the economic growth rate. In many developed countries, the IT industry has become the largest industry in the national economy.

"Perception China" is an image of China's development of IoT. The IoT allows objects to be sensed or controlled remotely within existing network infrastructure, creating opportunities for more direct integration of the physical world with computer-based systems. In addition to reducing human intervention, it improves efficiency, accuracy, and economic benefits. By implanting a variety of miniature induction chips on the object to make it intelligent, and then with the help of wireless networks, to achieve "dialogue" between people and objects, and "communication" between objects, have resulted in the birth of driverless cars. The IoT can show that anything in life can become "feeling and thinking." Such an intelligent picture is considered to be the world's next wave of IT and a new economic engine.

IoT is utilized in automated factories for producing food crops like rice, wheat, soybean, corn etc. as well as automated factories for raising pigs and cows. The Chinese Academy of Sciences, industries, and a number of universities have established an IoT Research institute in Wuxi, and Jiangnan University has also established the country's first physical IoT factory college. The IoT has been officially listed as one of the country's five emerging strategic industries. According to reports, the market size of the IoT in the fields of security, transportation, power, healthcare, and logistics has reached more than 400 billion yuan.

The market continues to grow, and today people have entered the era of big data and the digital economy. Big data will bring convenience, ease of life, social progress and environmental protection. Digital economy refers to an economic system in which digital technologies are widely used and thus bring about fundamental changes in the overall economic environment and economic activities. The digital economy is also a new socio-political and economic system in which information and business activities are digitized. Online transactions between businesses, consumers, and governments have grown rapidly.

The digital economy covers communication , computer technology , software , Internet , e-commerce, entertainment and medical service. Government agencies have the largest repositories of digital information. According to a report released by the China Academy of Information and Communications Technology, the value of China's digital economy reached 50.2 trillion yuan in 2022, accounting for 41.5 percent of national economic output, ranking second in the world after the US. By 2030, it is estimated that China's total digital economy may reach 99 trillion yuan, with an average annual growth rate of 16%, mainly driven by the digital industrial economy.

AI has been a hot topic lately, when Hangzhou based Liang Wenfeng released his open-source Deepseek AI in January 2025. His company claims that it will train its V3 model for US\$ 6 million, compared with US\$100 million for Open AI's GPT-4 in 2023. Deepseek's performance power and quality sent shockwaves through the world, causing AI leaders like Nvidia to suffer the largest single-company drop in US stock market history, with Nvidia losing US\$600 billion in market value. DeepSeek AI systems work by processing vast amounts of data, identifying patterns, and using algorithms to make predictions or decisions, essentially learning from experience to perform tasks that typically require human intelligence. AI applications are being used increasingly in many fields including industries, finance, agriculture, education, retail, healthcare, and manufacturing. More innovative applications will emerge as time goes on.

6.2 Digital Communications

The speed of communication has a huge impact on human progress. The faster the speed, the faster the economic progress. Digital communication is superior to paper media, oral communication or telegraph transmission. High-speed rail and expressways have contributed to the growth of the transport economy. The speed of communication is closely related to the following: daily life, government administration, domestic trade, financial transactions, banking, business operations, and geopolitics. 5G can help speed up data transmission. Many jobs can be done at home instead of in the office, as evidenced by the impact of Convid-19. Every day, digital information permeates our daily lives. Through digital information, we can get news from all over the world on a smartphone. Everyone can buy what they want online and then have it delivered to the door step. This online portal offers a wide range of products and varieties from many manufacturers. Now, farmers have access to digital information on how to grow their crops and the types of fertilizers and pesticides they should use. Farmers can buy fertilizers and pesticides on the website. Farmers can use selfie videos to sell their produce. Information on world prices can be obtained online for different grains such as wheat, soybeans, corn and pork, as well as the impact of severe weather such as droughts and floods on food production.

Using a smartphone, each person can book a taxi and travel to their destination at an agreed price, rather than waiting for a taxi on the street, as in the past when taxi drivers could overcharge passengers by driving longer routes. People can order takeaway food using their smartphones. It is possible to use mobile phones to make payments and transfer money. Paper money was invented in China in the 9th century and is now being made obsolete by digital payment. Chinese tourists traveling to other countries can carry a language translator with them to communicate with foreigners. Digital communication technology makes online learning of a foreign language easy. Digital communication enables a remote doctor to operate a patient located at a distant location.

We can work at home during the COVID-19 quarantine period. Everyone can use their smartphone to text, talk, or video each other. By installing a video camera in your home, you can use your smartphone to observe what is happening in your home. Each person's digital medical history is stored in a central national medical database, which any doctor in China can access and utilize for patient treatment. Using digital information and 5G equipment, doctors can perform critical procedures remotely, and this has revolutionized the healthcare system, benefiting medical professionals and all patients across the country. With this digital health system, many emergency patients will be treated quickly and appropriately. The government also saves a lot of money and effort by eliminating paper records and storage space. Digital communication is used in the following sectors:

- Government administration

Use of digital communications by the government is to achieve efficiency and productivity in a huge country with 56 ethnic groups, each with different cultures, languages, and traditions compared with the legacy method of using written paper. To manage the country effectively, the central government needs to send policy information to all parts of the country quickly. In the early 1950-60s, the Chinese government posted newsprints on notice boards in various city and town locations for citizens to read. Nowadays, government information can reach the public through various online media, and everyone can retrieve it with a smartphone. As a result, many people now do not buy newspapers to get news. The government can take immediate action to investigate citizen's reports on social media about any government official accused of corruption.

With digital communications, governments can quickly respond to any natural crisis, such as the 2019 COVID-19 pandemic, floods and droughts, and then take immediate action. China is considered a very safe place to walk on the streets at night, as many cameras are installed at important locations to monitor street conditions. If there is a robbery or a riot, the police will use facial recognition technology to quickly identify the people involved. Such a digital system saves a lot of human resources without the need to send police to do street patrol 24/7.

The Chinese government has taken a number of steps to make it possible for people to make inquiries online without going to a government office. In order to facilitate the public to pay taxes, taxpayers can pay taxes online. In the past, people had to spend time at the bank to get money and queue up at the government office to pay taxes. Such a traditional system wastes time and human and material resources. With Zoom, government officials in different places can have virtual meetings without having to travel long distances. The court uses the Zoom tool to allow remote witnesses to appear in court.

The government uses digital information and big data to provide enough passenger capacity on high-speed trains to meet the sudden increase in passenger demand during the holiday season. The government uses AI and big data to import sufficient wheat, soybeans, and pork from different countries in time to avoid food shortages caused by insufficient domestic production.

- Domestic market and foreign trade

As the domestic market and foreign trade situations are highly unstable, affecting the prices of oil, food and natural resources, instant market digital information reported on the local media will give companies an early signal to adjust their business plans. The digital information also can provide instant domestic market conditions that are affected by natural disasters such as severe inclement weather, earthquakes, and locust infestations which could affect food supply and prices, allowing the public to purchase food stocks in advance to avoid the crisis of food supply shortage.

In 2018, the US government significantly increased import tariffs on Chinese goods, thus affecting Chinese manufacturers, and Sino-US international trade declined sharply. Tariff details have been digitally communicated to all affected local manufacturers so that they can take immediate adjustment measures. Through social media like WeChat and Baidu, any daily news about fraudulent business practices affecting consumers can spread quickly and widely, which can affect the reputation of products and companies. This has also attracted the attention of the government to take action.

- Financial transactions.

Countries around the world use Swift to conduct international trade transactions and settle in US dollars. This is because the US dollar is used as an international trade currency and many goods are priced in US dollars, such as petrodollars. Swift is a digital messaging system that authorizes transactions for ready payment by the buyer to the seller. The final settlement is made through a mutually approved bank. When the US imposed sanctions on Russia's use of Swift in 2022, Russia launched a new digital settlement system using the ruble and digital yuan, bypassing US sanctions on the use of SWIFT.

Through public media, TV channels and electronic information on the internet, investors can see the trading volume of major stock markets at home and abroad in real time. Since the start of the war in Ukraine, the media reports daily changes in the international currency exchange rate and depreciation of the currencies of various countries. China starts using digital Yuan for cross-border trade. With digital communication systems, cross-border trade can flow smoothly. As China's international trade grows via BRICS, BRI , SCO and RCEP dealing with different national currencies, the Chinese government needs to expand and improve the digital settlement system.

- Banking business

Years ago, going to a bank to make a deposit, withdraw money or exchange money was time-consuming. Everyone needs to show their ID card, passbook and queue up. The banknotes were dirty and creased. If you are buying a 60,000 Yuan car, you need to spend a long time at the bank counting the money to make sure there are no counterfeit notes and the amount is correct. With the introduction of digital information on the internet, banking has become very convenient, and withdrawals and deposits can be made through ATMs. If a large amount of money needs to be transferred to another party, both parties can go to the bank to make the transfer without the need to count money. With the launch of Alipay and WeChat Pay, people can transfer money on their smartphones for payment to the vendor. With digital communication systems, bank operating costs are significantly reduced, customer service is improved, and banking efficiency is increased. According to the media report, Huawei is integrating the digital Yuan central bank digital currency (CBDC) into its Harmony OS NEXT operating system. The integration will make the digital currency more accessible and easier to use for all, up to 1 billion smartphone owners.

- Industrial Application

Many large industries use AI-controlled robots to produce products efficiently with high quality. With real time digital information, materials can be sourced real time from around the world at the cheapest price. Industries that focus on leveraging innovative technologies and digitalization can grow faster, reduce costs and improve their operations. With digital technology, Chinese EVs manufacturers can change their vehicle models very frequently to meet the changing needs of customers in various countries and to compete with competitors. EV manufacturers apply AI to collect real time data on global sales, customer demand, production capacity, manpower needs, and raw material supply chain information. Chinese EVs are favored by many foreign buyers around the world as all the Chinese EV have incorporated the state-of-the-art technology in the car design that is unavailable in the EVs produced by American, Japanese and European car manufacturers. China introduces the first driverless taxi (robotaxi) in the world, incorporating the BeiDou Navigation Satellite System. Chinese EVs incorporate a plethora of high-tech features like massaging seats, in-car karaoke, TikTok/WeChat integration, rotating touchscreens , refrigerators, self-parking remotely, 5G HD entertainment. The BYD SUV can wade through water up to 1 meter deep. Through digital information and cloud computing, many airlines collect data to identify optimal solutions for all operational requirements. This enables them to achieve the lowest costs and highest profit margins while providing the best service and convenience to meet passenger needs. In addition, the airline works with hotels around the world to provide room reservations and price lists on the airline reservation website. Car rentals are also available on the airline's website.

Before the digital economy, such mobility convenience did not exist. Buying tickets online 24/7 saves passengers time and provides convenience and makes airline operations more efficient and profitable. It took time and hassle in the pre-digital age when passengers had to go to an airline office to buy a ticket. In addition, for interline tickets, passengers had to wait many days for confirmation whilst the airlines used telegram to transmit information to other airlines . When the connecting flight changed time or date, the passenger had to return to the airline to change the tickets. Digital information improves business operation and convenience to customers.

- Geopolitics

The political competition between China and the US, and between Russia and the US, has led to increasingly complex geopolitics. Therefore, China needs a comprehensive communication system to collect and collate political news from around the world that affect China's security and trade in a timely manner. The frequent change of governments in Europe has had a huge impact on China, with some US allied politicians becoming very vocal and anti-China due to the influence of US political and money politics. The CIA persists in creating tensions in Hong Kong, Taiwan, and Xinjiang by spreading propaganda and misinformation. It uses money to bribe dissidents within China to cause internal civil disturbance. China needs the latest IT to keep track of dissidents and foreign spies and cyberattacks. China needs to use the latest high-performance software to play war games to find out where China's defense weaknesses lie, as well as where the US greatest strength lies and what chance China has of winning a war should it occur.

The large number of American troops and warships abroad is constantly being moved around to meet the ever-changing geopolitics worldwide and a fresh US political strategy with a new US President. American politics are very unpredictable and aggressive as American foreign policy can change suddenly to meet the intrigue relations between the Congress and the Wall Street elites, the military industrial complex and powerful oligarchs in the US financial world, the intrigue relation between the President and the deep state, and the influence of the congress lobbies . It is an open secret that many Congress senators are multi-millionaires. Many senators in Congress, some over 75 years old, have held their positions for durations exceeding those of many dictators in other countries. For instance, Robert Byrd (1917-2010) served as a senator for 51 years, holding the longest senatorial position until his death at 93.

- Blockchain

Blockchain is a very safe digital ledger for recording information that is difficult to amend. Blockchain is used to reconcile legal contracts that are secure, self-executing digital contracts. It can automatically execute all or parts of an agreement with a vendor when conditions have been fulfilled. Blockchain also provides a secure method of storing records and exchanging documents between two parties. Blockchain is used in the hospital patients' records, manufacturing, supply chain, media, education, government, cybersecurity, agriculture, infrastructure, and energy. Blockchain technology can automate and digitize processes for boosting efficiency and speed. Blockchain is the technology that underpins Bitcoin, and it was developed specifically for Bitcoin, which is a decentralized digital currency.

6.3 Digital Medical Services

With the advent of digital imaging, the need for many full-time radiologists working in a hospital to interpret medical images has been greatly reduced, as digital images can be interpreted by any qualified radiologist located in another hospital. Compared to conventional X-rays, digital X-rays provide better and sharper images with high resolution that can be enlarged or manipulated as needed, minimizing the need for repetitive operations. Because conventional X-rays are printed on film, they must be kept in a safe place. Digital imaging requires less radiation to produce images of the same quality as film. The digital images produced by most medical imaging equipment (CT, magnetic resonance imaging, ultrasound, nuclear medicine, and routine radiography) can be sent via internet networks to the doctor that requires the image. The advantage of digital imaging over film-based radiography is that the spatial resolution of digital radiography is of higher magnitude than that of screen film radiography.

Eric Topol, an American cardiologist, scientist, and author, is the founder and director of the Scripps Research Translational Institute. He said that China had obvious advantages in the application of AI in the medical and health field, and was in a leading position in the world, including the use of WeChat to pre-diagnose patients, collect patient data, and analyze CT scan results. Instead of confrontation and competition, the US and China should cooperate more on AI development. Topol said: "The US needs to continuously improve its data collection capabilities in clinical care to catch up with China's pace of development." If the US government pursues a policy of suppressing and isolating Chinese technology companies, it will greatly harm the interests of US companies and hinder their progress in big data technology in the healthcare sector. He said that China is at the forefront of the world in facial recognition, smart city management, mobile payments and other technologies.

6.4 Unmanned Aircraft

Not all wars are won by attacking the enemy. With a good military defense system, you can win some wars. In addition, a good military defense system may be cheaper than an attack system. On 7 May 2025, Pakistan advanced radar defense system helped shoot down 3 Indian Rafale fighter jets by Pakistan's China-made J-10C fighter jets. The destroyed jets fell down in Indian territories. Rafale jets cost more than 6 times that of J-10C. Currently, the US military strategy relies on its massive military attack strategy operating from 750 military bases around the world. And the US armaments on land are mainly old weapons left over from World War II. Its military operations rely mainly on a large naval force that carries out air strikes on enemy territory.

The history of past wars shows that American ground troops have not achieved anything fighting on foreign soil, such as North Korea, Vietnam, Iraq, Syria, and Afghanistan. Since China is a nuclear-armed power, the US would not dare to attack China as this constitutes a declaration of war. Even the US dares not attack the nuclear-armed Soviet Union or Russia for over 70 years. As long as China has a good defense system, the US cannot win a war with China in the South China Sea. American military planners played a lot of computer war games, and out of 10 games, the US did not win a single war with China in the South China Sea.

China produces many intelligent digitally operating drones for various commercial and military applications, even the Ukraine army uses Chinese DJI made drones to attack Russia. Using swarms of unmanned aircraft and IT drones fired from the shore to overwhelm the US navy ships in South China Sea, the US has no chance to win a naval war with China. IT drone and unmanned aircraft swarm warfare has the following advantages:

- Low cost, positive effect.
- Reduce combat casualties.
- Part of the "swarm" can be sacrificed to act as a decoy or jammer to entice
- enemy air defense detection equipment to turn on and expose its position.
- Inducing enemy's expensive anti-aircraft missiles firing at cheap drones and
- unmanned aircraft can quickly exhaust enemy stocks
- China can use undersea smart drones to attack US navy ships, particularly the ship's propeller, thereby crippling the ship.

According to combat mission deployment, the drone "swarm" can compose a variety of modules, such as reconnaissance and detection, information processing, fire strike, and so on. Finally, the drone "swarm" can also cooperate with the man-machine formation, manned aircraft from the rear to command and control the drone "swarm," so that it can conduct reconnaissance and strike in complex and unknown areas, reducing the risk of combatants. Moreover, the cooperation of the drone "swarm" and the man-machine formation makes strike range more concentrated. In the future, unmanned aerial vehicle UAV "swarm" combat technology could enter a new era of comprehensive development, which may involve cross-integration and comprehensive application of various related technologies such as AI, quantum technology, semiconductors, land based radar and autonomous systems. American experts say that China's "unmanned swarm" tactics are growing into a new generation of "aircraft carrier killers." In the South China Sea

NOTE: In April 2025, Houthis in Yemen fired drones at US aircraft carrier USS Harry Truman in the Red Sea, resulting in two F-18 fighter jets each worth $67 million falling overboard. It happened when the aircraft carrier was making maneuvers to avoid Houthi assaults. The Houthis is not a national regular army but a militia. In May 2025, the US reported the Houthis almost shot down an F 35 fighter jet in the Red Sea. In 2025 US Defense Secretary Pete Hegseth's wife reportedly attended meetings with foreign defense officials with sensitive information being discussed. He was also criticized for inviting a journalist in a chat group discussing secret information on a military strike on Yemen. He did not even know what ASEAN stands for when he was questioned by the Senate hearing. How can the US fight a hot war with either Russia and China when the senior US politician is such an ignoramus.

China reportedly has the largest aircraft cemetery in Asia and the second largest in the world, at Lushan Military Airport in Pingdingshan City, Henan Province, which houses more than 2,000 aircraft, including Shenyang J-5, J-6 and J-8 fighter jets. China can convert retired fighter jets into UAV, which can be used as decoys to help carry out suppression of enemy defense systems. Using current AI and autopilot technology, the retired jets could be given a new lease of life by playing a role in a new military combat operation.

It is a good defense and economically viable strategy to use a fleet of 'Kamikaze' drones made up of 100-200 retired fighter jets, converted into drones carrying powerful explosives to attack and sink American carriers in the South China Sea at night. The average cost of building a new aircraft carrier is between $13-15 billion. The US aircraft carrier has 80-100 fighter jets and 5,000 personnel, including pilots, onboard. The US relying on naval vessels far from its home base to attack China's powerful land-based military in the South China Sea is destined to be a failed mission. Drones using AI systems can accurately attack target objects using the BeiDou global navigation satellite system. American supply warships are slow-moving objects and are a good target for Chinese supersonic missiles to attack. The US still relies on an old-fashioned single-base radar system onboard the aircraft carrier, whereas China can use an advanced integrated radar system on land and AWACS and BeiDou satellite global navigation system.

6.5 Aircraft Digital Design

Boeing, Airbus, or Comac does not manufacture all the parts, components, and accessories needed to build an aircraft. They source many parts from OEMs and subcontractors from various companies, including domestic and overseas subcontractors. The entire process of manufacturing an aircraft is supported by the material supply chain. The aircraft is built with a variety of electrical wires, air conditioning system, fuel system, hydraulic system, air pressure system, fire detection system, electrical power system, power plant and landing gears on the aircraft and are designed and provided by external suppliers, OEMs. Aircraft manufacturers assemble all the parts and components according to schedule. All the suppliers production schedules must dovetail with the aircraft manufacturer's schedule via the digital information system. The advanced aircraft integrated design system ensures that every part and component fits perfectly within a very tight margin of tolerance.

Comac is new in the business of building commercial aircraft as compared with Boeing or Airbus who have extensive experience in building commercial aircraft since WW II. It has built up a seamless supply chain of suppliers of materials and subcontractors and a state-of-the art digital database of aircraft commercial operating experience and technical knowledge. Boeing also manufactures military hardware for the US government and hence receives funding for R&D and purchases of supercomputer systems which have dual application for commercial and military application. Therefore, Comac needs funding from the Chinese government to purchase supercomputers and servers to perform aircraft and system digital design projects and also integrate the digital design of the supplier's equipment and operating parameters.

The design and manufacture of commercial aircraft involves many complex processes and operations, given the constraints of many factors such as aircraft weight, flight safety and customer operating costs. Boeing accumulates massive data from wind tunnel testings of various aircraft scale-models of wing airfoil. The results of these tests are then used to optimize the model's wing design for various aircraft configurations required by the airline customers. These data are stored in the cloud computer. Wing designers can interpolate previous wing tunnel testing data for new wing aerodynamic profile for new aircraft models. Such big digital data storage is of immense help to speed up new aircraft model design.

Boeing has accumulated a wealth of data from past designs and manufacturing, from different suppliers and costing, and from different customer requirements. Each new aircraft delivered has a different interior configuration. So Boeing uses supercomputers to cater to the demands of various customers around the world. Boeing has stored aircraft structural repair information in the supercomputer so that airlines can download the latest repair information instantly. Previously, the repair information was stored in microfilms that are kept in various airline technical libraries.

6.6 Digitalization of Integrated Industry

A successful integrated industry needs to implement digital information and communication systems to operate seamlessly. During the pandemic, the auto industry was severely affected by a shortage of chips due to the lockdown of chip factories. The automotive industry is a highly integrated industry that requires different companies to supply different parts at once. The newly established Apple company in the US was interrupted by a local shortage of small screws, which later had to be imported from China. The local shortage is due to suppliers' inability to increase production to meet demand. A broken supply chain affects all digitalized integrated industries. The strength of a supply chain depends on the weakest link.

Alibaba CEO Ma Yun, like Steve Jobs, is an industry integrator. The two companies bring different products together to form a larger digitalized integrated industry that appeals to a wide range of consumers. Boeing is also a digitalized integration company, assembling many products from many companies around the world to make one airplane. Boeing cannot design and manufacture all the components that are installed on the aircraft, such as engines and landing gear.

In the driverless EV Robotaxi service, a new type of integrator emerges, utilizing 5G communications, AI and IoT, and multiple sensors. This integrator takes advantage of digitalization introducing a new industry that eliminates the hassle of owning a car, the burden of maintenance and repairs, sudden mechanical breakdown, annual depreciation, paying insurance and parking fees. Drunk driving and traffic offenses are avoided. In addition, Robotaxi EV service does not require many private car park spaces in office buildings or factories. This driverless EV is safe for the young, the elderly and the incapacitated to travel by day or by night 24/7.

China has built one of the world's largest online shopping platforms to sell products made in China. The digital platform is available in different languages and contains detailed information on each product. Online sales of electric blankets on Alibaba's cross-border e-commerce platform in Europe have more than tripled due to US sanctions on Russian natural gas exports to EU countries in 2022. Gas is used to heat the EU country's homes in winter. Americans buy many farm tractors and small utility trucks online through Alibaba's platforms. Another product that American boat owners often buy online is a China-made outboard engine. Alibaba has integrated digital payment and logistic systems involving product suppliers and customers.

Other digital industries that China can embark on the following projects:

- Transforming an old city to a smart city using IoT, cloud computing, big data, spatial and geographic digital information integration and next-generation information technologies to promote digital urban planning, management and social services. The construction of smart cities is of great significance to accelerating the integration of digital industrialization, IT application, urbanization, and agricultural modernization. It also enhances the ability of urban sustainable development to meet the demand of the modern digital world, enabling the people to further improve their living standard and quality of life.

- China can carry out turnkey projects in fully integrated digitalized hospital design and construction for export to Africa, Central

Asia, the Middle East, South America and Asia. These projects include hospital equipment, instruments, information technology systems for medical management, operating theater, laboratories, CT scan and MRI facilities tailored to suit each client. China can provide training to the client's hospital operation staff and management onsite and in China.

- China's new AI agriculture farm is an integrated digital industry that includes planting, growing, harvesting, packaging, storage, transportation and ultimately distribution to consumers. The entire process is controlled by AI and supercomputers. With these AI agriculture farms, China will achieve self-sufficiency of grains. This AI integrated digital farming industry can be built in other countries to produce the crops they need.

- China is to set up the world's largest EV rental company with branches around the world. The company is to be managed by China's largest EV producer BYD, using BYD built EVs with the cheapest price in the industry. Comac can also set up a C919 aircraft leasing company to compete with Boeing and Airbus. Leasing aircraft allows airlines to reduce initial capital investment. Comac may also offer leased aircraft maintenance, repair work, parts, and accessory replacement support. Aircraft leasing business will enhance sales of Comac aircraft.

6.7 ROBOT CHEFS AND FULLY INTEGRATED RESTAURANTS

T he government endeavors to retain the traditional age-old cuisine, method of cooking, composition of ingredients from various parts of the country. A state-owned cuisine company is to be established to employ all the experienced chefs as consultants to record the entire process of preparation of each dish so that a robot chef can replicate the real dish. There will be as many robot chefs as many dishes to be replicated.

During the 2022 Winter Olympic Game in Beijing, the government set up smart restaurants at the Olympic Village and media center which had fully automated kitchens where robots prepared and served meals, including Chinese dishes, burgers, fries, and cocktails. These robotic systems used robotic arms and conveyor belts for cooking and delivery. According to media coverage, reports highlighted robotic chefs frying rice, noodles, sandwiches, flipping burgers, making ice cream and coffee. Chinese state media noted over 120 types of robots deployed across the Games for roles like cleaning, delivery, and cooking. Dozens of robots were used for food preparation.

Every dish produced by each of the robot chefs is to be verified by an experienced chef to match the original taste and is to be assigned a number and a dish name. Each dish will be patented. The company is to be named as Chinese Robotreat (CR) like McDonald and will franchise its business to private restaurant investors. CR will provide robots to the franchisee who will decide whatever dishes he wants to offer to customers. CR will control the supply and maintain quality of all ingredients. The operation of each franchisee restaurant will be monitored digitally via a daily online report of sales and complaints.

Each franchisee restaurant will be manned by robots. If all franchisees collectively sell 100 million dishes daily across the country to students, office and factory workers, family members, and tourists, and each dish yields a net profit margin of 10 Yuan, the total daily profit could reach 1 billion Yuan, or 365 billion Yuan annually. When CR is successful after a few years of domestic operation, it can franchise the business to foreigners in other countries to compete with McDonald , Burger King and KFC. It is to be noted that Chinese robotic restaurants can operate 24/7.

In China, many housewives wake up very early in the morning to prepare breakfast for their children before going to school and also for their husbands before going to work. When the breakfast is over, the housewives need to wash the dishes and cooking utensils. They then follow by washing clothes and mop the floor. Their chores can be compounded by their aged parents, who are too old to be independent to feed themselves. The housewives may need to buy different food ingredients to prepare special foods for them to eat every day. Hence, it is very tiring for the housewives to do their daily chores. With robot chefs restaurants in the neighborhood serving a variety of meals and providing food delivery to the doorsteps, the lives of the housewives will be less burdensome and pressured. Their lives will be less stressful. The robot chef restaurants are a blessing to those single elderly people living alone without domestic help to cook for them. This is very helpful to them during cold winter and hot summer times.

6.8 WORLD'S LARGEST DIGITAL BESPOKE DRESS MAKING ENTERPRISE

Clothes are necessities in everyday life. Hence, dress making will always be an evergreen industry. Many people, especially women, go to the shopping malls and have difficulty finding their desired style, color and size dresses. It is also troublesome to find an experienced tailor to make a bespoke dress, pants, or shirt to suit oneself. Using the latest technology, a private company is setting up the world's largest integrated digital bespoke clothes making company in China. The scope of this company's business includes: the use of robots to measure the three-dimensional figure of the customer body to make a bespoke dress for the customer who can select from the online catalogues of various designs and styles of dress, pants, or shirt. The customers can also select the color, fabric, and buttons.

The final selected dress, pants, or shirt outfit will be displayed on the company website for review and acceptance. Once the 3-D physical dimension is taken, it will be given a code for future dress making reference. Fashion designers in any country can upload their designs to the website for a design fee. This new integrated digital bespoke dress making is good for male and females of various body sizes. The dress making factory operates 24/7 by robot workers. The final dress will be delivered to the customer residence.

One entrepreneur establishes a high fashion bespoke woman dress design and making company located in shopping malls around China. The company can take the customer 3-D physical dimension. The company will use AI to make several suggestions of fashion design. Color, fabric, style, to suit the customer requirements to dovetails her age, profession, social status, special social occasions and her budget. She can view the various dress designs on a large computer TV screen with her image and various hairstyles created by a generative AI deepfake image. The dress design can be altered to meet her final requirement. Various costume jewelry to suit the various dress designs can be added on the computer screen. The whole shop setup is similar to a movie set dressing room where the actress is dressed up by a professional costume designer. This is the world's first generative AI designed bespoke woman dress.

In the factory, robots will use a computer-aided robot cutting machine in accordance with the design of the dress pattern and the customer's 3-D physical dimension. It automatically cuts the cloth to several pieces to precise dimension with a minimal wastage of cloth. The cloth edges are then glued with special glue paste to make the final bespoke dress. This process eliminates the use of a traditional sewing machine. The robots can make bespoke dresses or pants in a matter of minutes.

This integrated one-stop dress making company has long-term contracts with fabric materials suppliers around the world. This robot dress making factory can be established in any country around the world, providing a high quality and fashion bespoke woman dress for various occasions like wedding, dancing, partying and special events for women customers. It is a new integrated revolutionary industry involving various global supplies of fabric and raw materials, designs, software and new technology.

6.9 QUANTUM COMPUTING

In 2016, China launched the world's first quantum satellite using cutting-edge laser technology. The navigation satellites are equipped with laser communication equipment that can transmit data to the ground at a rate of several gigabytes per second, a capability that the US GPS does not possess. According to a report, on June 16, 2021, the Beijing-3A satellite was launched to scan the San Francisco Bay Area. In an experiment, the satellite performed an in-depth scan of the city, covering 3,800 km² in only 42 seconds, with enough clarity to identify a military vehicle on the street and determine the type of weapons it carried.

In February 2023, the US sent a fighter jet to fire two missiles that shot down a slow-moving weather balloon drifting into US airspace from China. The US claimed the weather balloons were spy balloons used to photograph US land objects. The weather balloon was so large that it could be seen with the naked eye from the ground. The US forgets that China can take better pictures with a small satellite than a balloon camera. In response to the hysterical public reaction to the Chinese weather balloon, the US Air Force fired $2 million worth of missiles to blow up the $12 balloon. No wonder the US government has to increase its defense budget every year.

According to a report, on May 3, 2017, the Center of Excellence for Quantum Information and Quantum Physics of the Chinese Academy of Sciences announced that China had built the world's first quantum computer based on single photons, a development that surpassed earlier classical computers and could upend traditional computing. China's quantum computing research is getting closer to making the crucial shift from quantitative to qualitative change. The new quantum computer can beat Jiuzhang, the most powerful supercomputer in a competition that the quantum computer could complete a calculation in 200 seconds that would take a supercomputer 2.5 billion years to complete. Faster speed of transmitting or processing digital information helps China progressing faster in every aspect of scientific and technological advancement. The quantum satellite will propel China to greater heights in the field of AI. More innovative digital products will be produced in China faster than other countries.

Chapter 7 Internet Of Things (IoT)
7.1. IoT Technology for Digital Economy

The IoT is a network of physical objects or things embedded with sensors , software, and other technologies that allow them to connect and exchange data with other devices and systems over the internet. Such is the world's new information technology for expanded economic innovation. The IoT is officially listed as one of China's five emerging strategic industries. According to reports, the market size of the IoT in the fields of security, transportation, power, healthcare, and logistics has reached more than 400 billion yuan. The market continues to grow in new industries.The new electric vehicle makes use of IoT to enable it to be driverless.

Human progress is closely related to speed. China's high-speed railway construction in 10 years has reached the world's longest 45,000 km railway. High-speed rail with speed of 350 km/h has made use of IT and IoT with improved safety standards. High-speed accelerates economic progress. The world today is undergoing the most rapid, extensive and profound changes in human history with ever-increasing new innovations and technologies. At present, the competition in the information industry between China and the US has reached a peak. The profound impact of informatization on economic development and social progress has aroused widespread concern around the world. IT and related industries are growing 2–3 times faster than the traditional industrial economy. Xiaomi is building a smart automatic robotic factory without a human worker to make potentially 60 smartphones in one minute. This factory leverages IoT technology for real-time data monitoring and machine-to-machine communication. This is one of the reasons that China's economy has astonishingly progressed so rapidly over the past 70 years.

7.2 IOT APPLICATIONS

The IoT, a product of Internet development, analyzes and processes vast amounts of online data. This data is then transmitted to service providers through big data mining and predictive processing, thereby expanding the application of AI. The IoT is another IT revolution that has arisen in the world. The industrial value brought by the IoT technology is many times that of the IT. With 6G in the future, the transmission of cloud computing in the IoT will be more than 1,000 times faster than 5G, with instantaneous effects on telemedicine, driverless cars and power grids. IoT plays an important role in military operations. Application of IoT also improves the quality of life, reducing the waste of natural resources, increasing the efficiency of urban planning and infrastructure maintenance.

The industrial IoT platform can provide customized services for the individual needs of large enterprises, helping them reduce costs, increase efficiency, and improve decision-making through intelligent analysis of data collected from equipment, production lines, and other operational links. Automotive sensor-driven analytics and robotics improve the efficiency of vehicle manufacturing and repair. For example, industrial sensors are used to provide 3D real-time images of components inside a car, allowing faster diagnosis and troubleshooting when replacement parts are automatically ordered by IoT systems. Logistics, land and marine transportation scope, commercial and industrial IoT devices can help with supply chain management, including inventory management, supplier relationships, fleet management, and scheduled maintenance. Shipping companies use industrial IoT applications to track assets and optimize fuel consumption on routes.

Convenience stores using IoT technology do not need manual workers. Customers can scan the QR code to open the entrance door, pick up the merchandise from the shelves, and pay for the purchase at the cashier station by scanning each merchandise. In an AI-enabled IoT smart car parking management system, a remote camera can detect the car license plate which has a sensor embedded, and when the car exits the car park, the car owner will be charged for parking fees automatically. Shenzhen bus company uses AI and IoT technology to monitor its many thousands of buses operating on the roads. The system can pinpoint exactly the location of every bus at any time and also road conditions.

The Global IoT Spending Guide predicts that by 2026, the manufacturing industry represented by smart factories, the government projects represented by smart cities, the retail industry represented by online operations, and the public utility industry, represented by smart electric grids, will account for more than 60% of China's enterprise-level IoT market spending. Some experts analyze the progress of the IoT that by 2030, the global industrial IoT projects can create 14 trillion US dollars of output value. China has become the world's largest user of IoT.

7.3 CONTRIBUTION OF IoT TO AGRICULTURAL PRODUCTION

China has insufficient grain production and needs to import a large amount of corn, wheat, and soybeans every year. Import countries include two unfriendly and unreliable countries, America and Australia. One day, these two countries might sanction China and ban the export of grains to China. Food shortage will cause civil unrest. According to a report, China imported 138 million tons of grain in 2024, including corn worth $6.7 billion, wheat worth $4.5 billion and soybeans worth $58.5 billion, with a total value of $69.7 billion. In this case, China will lose $69.7billion annually to foreign countries forever.

China can save this amount of money annually by investing in domestic grain production, leading to accelerating China's economy and providing jobs for tens of thousands of people. The annual amount of $69.7 billion will recirculate forever in the country's economy. AI and IoT technology automatic agricultural farms can be set up across the country to grow wheat, soybeans, corn and other grains to overcome food shortages. The automatic farms can produce all kinds of grains required for national consumption as well as for exports. The farms can be located across the whole country to reduce cost of transportation.

The IoT and AI farm is environmentally friendly. Plants are grown, processed, packaged, stored, finally transported and distributed to consumers in every city in accordance with the latest information generated by the cloud computer big data. Information on daily grain consumption in each region is transmitted to the central State Food Administration for planning food production. The entire operation uses AI technology to produce the required quantity of grains for each region. This closed-loop AI control system ensures that the supply chain is flawless and waste-free, thus maintaining the lowest cost to the consumer and keeping food prices at a stable level. This automatic robotic farm industry has an important contribution to China's socio-economic progress, national security and GDP.

7.4 IoT Solve Power Shortages Problems

Every region in the country has electric grid network facilities. The grid has three roles to perform: to ensure the optimal use of power from power stations; to provide greater supply capacity, and to make the power system operate more economically and supply backup power. The power grid is an interconnected network that provides distribution of electrical energy from power plants to each consumer. Interconnecting power stations to each other is to allow the spare generating capacity as a rotating energy backup. The grid is defined as the network connecting the power generation, transmission and distribution units, which provide electricity from the generator set to the distribution unit, and a large amount of electricity from the power plant to the load center. Electricity generated is wasted if it is not used.

The importance of the grid lies in the ability to instantly supply backup power to a region with power shortage. The distribution system is balanced whenever power supply and demand reach synchronous mode. However, the old manually operated power backup grid system cannot immediately cope with the need for emergency power delivery. Consumer demand and consumption changes from day to day and power stations are subject to severe rainstorms, floods and earthquakes, resulting in power outages in the affected regions. China is a vast country with unpredictable power shortages every year. Prolonged power outages in big cities will seriously affect industrial operations, human activities, traffic paralysis, and a huge economic loss.

IoT digital grid systems can improve power supply reliability and reduce power outages. Special meters and sensors on transmission lines installed in homes and businesses can continuously monitor demand and supply, while mailbox-sized devices called synchronous phase-fluxes can measure the flow of electricity through the grid in real time, enabling operators to anticipate and avoid outages. IoT can collect all the data from power stations and consumers' electricity use in real time. It can also automatically connect the power supply of each power station to maintain power distribution balance.

7.5 THE CONTRIBUTION OF IoT TO HEALTHCARE

The human body is complicated. There are four symptoms that make a person feel sick: pain, swelling, skin discoloration, and high body temperature. During the patient's first hospital visit, the most difficult consultation is for the doctor to obtain an accurate description of the illness from the patient. Sometimes doctors feel too tired at the end of the day to ask patients all the necessary questions due to the physical exhaustion of treating too many patients in one day. Some patients are afraid to go to the hospital because of the long wait time, especially from a distant place, spending a long time on the road and waiting too long at the hospital. It's inconvenient to go home too late. The medical equipment of small city hospitals is not complete, and it needs the remote assistance of various specialists in big city hospitals.

The application of IoT will improve hospital care for every patient. Before a patient arrives at a hospital for treatment, he must fill out a questionnaire form on the government hospital website, including his personal information, illness symptoms as well as any genetic diseases in his family, such as cancer, diabetes, and hypertension and questions on smoking, drinking, and exercise. The patient must answer all the questions before going to the hospital for treatment. The patient will receive a message to tell him which hospital to go to and time of appointment. The patient's medical record is kept by the government's health ministry.

Upon arrival at the hospital, he may need blood tests or X-rays. He will be told the name of the doctor to see. If the patient's condition is mild, he will be directed to the outpatient clinic for a consultation. If the patient speaks only the local dialect, he will be accompanied by a nurse who is familiar with the local dialect. When the patient arrives at the doctor's office, the doctor will have all the medical conditions of the patient ready. The whole process may only take a few minutes. If the patient needs a physical CT scan, he will be directed to the CT scan room. After the CT scan is completed, the patient needs to return to the doctor's office for final consultation. If a patient needs immediate surgery, such as heart disease treatment, he will be directed to a surgeon, who will take him to the operating room.

IoT and AI technology will improve hospital operations and quality of service. For patients, the IoT technology will provide patients with better medical services, cheaper costs, faster consultations, efficient medical treatment, reduced waiting time. For hospitals, by recording a complete medical history, IoT technology can reduce stress on doctors and nurses, and improve the relationship with patients. X-ray and scan results will be analyzed by a computerized system of AI technology. The hospital questionnaire will provide a better allocation of specialists for patients with specific medical conditions. This is important, especially when dealing with patients who require surgery.

Operating room occupancy varies from hospital to hospital. Some hospitals' operating rooms may be fully occupied for a particular day, whilst others have empty rooms. The IoT system will show the operating room occupancy of each hospital on a daily basis to all surgeons in the city. Through remote surgery with 5G technology, senior surgeons in large hospitals can remotely assist hospitals in smaller cities, which do not have enough experienced surgeons to operate on a patient with complicated medical conditions.

With IoT collecting all personal medical records of cancer patients, university medical students can use quantum computers and AI like DeepSeek to analyze the root causes of cancer in a real clinical study. Difficult medical cases can be assessed by the medical university professor to let students learn how to treat the patient. The same procedure can be used to search for the root causes of other diseases, such as diabetes, dementia, and heart disease. Some American medical experts have commented that China possesses clear advantages and a leading global position in the application of AI technology within the medical and health field, including the use of online submission for patient medical histories. AI is widely deployed across Chinese tertiary hospitals, enhancing diagnostic accuracy, streamlining workflows and improving patient management. AI-powered pathology, imaging analysis and clinical decision support systems have demonstrated significant potential in optimizing medical processes and reducing the cognitive burden on healthcare professionals.

China has made significant strides in integrating AI into healthcare. In 2023, researchers at Tsinghua University announced a physical "AI-run hospital" treating real patients without human oversight. This project, described as an "Agent Hospital," is a virtual medical environment where AI-powered agents simulate doctors, nurses, and patients to train and refine AI systems for real-world healthcare applications. It is a simulated hospital ecosystem where AI "agents" role-play as doctors diagnosing AI-simulated patients. Designed to accelerate the development of medical AI by allowing the systems to practice diagnosing more than 10,000 potential diseases and interacting with virtual patients.

The AI hospital is not a replacement for human staff, but a training ground for AI to improve accuracy and decision-making. The AI doctors analyze symptoms, medical histories, and test results from virtual patients (also AI agents). The system learns from feedback loops, refining diagnostic and treatment recommendations. Its focussed areas include rare diseases, complex cases, and emergency scenarios. The goal of this AI-run hospital is to enhance diagnostic accuracy and speed and to reduce errors in high-pressure medical environments.

7.6 USE OF IOT IN AIRLINE OPERATION

IoT helps airline operations by enabling real-time data collection and analyses resulting in enhanced airline operation, improved air traffic management, enhanced passenger services, predictive airplane maintenance, monitoring various aspects of aircraft operation, engine performance monitoring and baggage location.

- Enhanced passenger service
- IoT–connected devices can provide personalized in-flight entertainment, optimize cabin comfort, streamline check-in and aircraft boarding
- Enhanced baggage handling
- IoT-enabled tracking systems can assist locate lost or misplaced baggage
- Aircraft fuel efficiency
- Real-time data analysis assists airlines optimizing flight paths and fuel consumption, resulting in fuel saving
- Aircraft operation safety
- IoT sensors can monitor various aircraft systems, providing early warnings of potential problems and contributing towards fight safety
- To maximize the passenger and cargo load factor, IoT can also assist airlines to optimize the number of crew changes in overseas stations. Airlines can also partner with hotels to provide convenient hotel booking at discounted rates and airport transfers. Previously, these tedious information collection and analysis tasks were performed by a group of clerical staff. Their

work was monotonous and human error was bound to occur. Every passenger can now book a flight online at any time of the day or night anywhere 24/7. The airline has the personal details of all passengers to prepare cabin service for all passengers, who can also print their boarding passes on the website. Technology reduces costs of operation and improves living standards.

Once an airline has developed the optimal flight schedule for its entire network, engineering maintenance can plan the appropriate schedule for each aircraft in the fleet, maximizing the flight time for every aircraft. The engineering department can gather daily information on each aircraft delay across the network, study the causes of engineering delays, and use IoT to take corrective action to avoid similar delays in the future. The engineering department must avoid aircraft being delayed beyond a certain time limit, which could result in a replacement of the crew. Engineering must be aware not to cause avoidable long delays that the passengers need to disembark. Line station ground engineers can overkill a stubborn recurrent technical defect by replacing more components to avoid long downtime in troubleshooting. Long aircraft delays can affect both upstream and downstream aircraft operations.

When each defective aircraft component is replaced at a line station, the information is collected and stored in the IoT database for AI analysis. The component's part number and its history, and the service life of the component part is then calculated. Based on big data, IoT will suggest replacing such components at a time before it fails prematurely. IoT technology can keep and monitor the history of all spare components stored at each line station. The system will prompt a particular component for timely update or modification. IoT can track the safe operation of all critical components (those outside the

Minimum Equipment List) to prevent premature failure during aircraft operation.

7.7 Use of IoT in Aircraft Engine Overhaul

The engine is the core component of the aircraft. Engine maintenance and overhaul are the most expensive items for aircraft operation. Using facial image computers in the workshop to detect defects and physical dimensions of engine compressor and turbine blades and vanes eliminates the need for manual inspections, which could lead to human error, and therefore saves airlines significant amounts of money. In addition, facial recognition can be performed 24/7, which speeds up turnaround time for engine overhauls.

A huge amount of data is generated from facial images of all the defective parts. Using this big data, parts manufacturers can research and redesign parts to improve the parts' reliability. It also allows OEMs (original equipment manufacturers) to more accurately plan how many spare parts to be produced to meet the demands of customers. This is important because OEMs need to buy raw materials, such as chromium, nickel and cobalt, for making engine turbine blades. These minerals are not readily available on the market and have a long lead time, especially when the current geopolitics affects the availability of the minerals in the market.

Using AI cloud computers to analyze the life span of various parts of the engine and the associated maintenance costs, airlines can establish a proper plan to dismantle the engine for overhaul at the right time and at minimal cost. IoT and AI will provide an excellent opportunity for airlines to improve their operational reliability and profitability while reducing labor costs. However, implementing IoT and AI systems require heavy capital investment, which may not be affordable to small airlines.

Chapter 8: China New Smart City
8.1 New Smart Government

Economics and politics are inextricably linked in a country. For China to advance with time and rapid urbanization, the Chinese central government must have strategic long term plans to guide industrial development to achieve economic growth. It is remarkable that China has achieved the second-largest economy in the world in a span of only 70 years. Now, Chinese cities have faced many social and technical problems such as water resources, pollution, housing, and congestion. These problems can be solved by new smart city design. The new smart city is the most realistic application and implementation of IoT and AI technology, coupled with smart factories and smart farming production in agriculture and livestock, enabling China to achieve self-sufficiency in foods and energy. The new smart city concept gives the new e-government a way to give enterprises the opportunity to adopt IoT and AI; to improve living and social conditions, citizens' health and national economic advancement. The new smart city can effectively integrate and utilize the city's resources; and further improve the city's production efficiency and service capacity. China has all the necessary resources to achieve its smart city goals

New smart cities will have the ability to collect information for transmission, processing, and interaction quickly and provide a self-sustained abode with sufficient renewable energy, food grains, fish, and meat. Education and healthcare are free. Housing is affordable for all newly married couples due to a new land reform not subject to land and housing speculations. Smart city living is immersed in a high-tech digital environment. Innovation in social infrastructure, education, and healthcare must keep abreast with time. Chinese Deepseek AI introduction in 2025 was a showstopper in the technical world, causing a phenomenal Nvidia market value of $300 billion erased in a day. With increasing productivity and efficiency in the government and industries, Chinese may need to work only 5 hours a day to sustain a good living, instead of the current 8-hour daily work.

As the US has been actively wanting to halt Chinese economic progress by all kinds of sanctions and tariffs relentlessly, thereby hurting Chinese trade with the US and the world, China's economy has slowed down and unemployment is rising. Hence, this is an opportune time for China to build 200 new smart cities in a progressive manner so as not to cause a sudden shortage of materials and other resources . It is expected that the construction of new smart cities will achieve remarkable economic benefits eventually.

These smart cities will lift the poverty of the remaining 600 million poor rural families. They will be educated and trained with appropriate skill to work in high value-add jobs in the supply chain. Skilled human resources are a cornerstone towards industrialization and economic advancement. China's human capital investment must continue to be prioritized as the ultimate driver of economic progress. With 600 million people entering the middle class, China's domestic market will be large enough to propel China to be the world's largest economy and the highest GDP, surpassing the USA. Human resource is like a goldmine that needs to be mined and processed to be useful.

Under the Chinese government's continuous drive of technological innovation, the construction of new smart cities will also move to a higher level of continuous economic advancement in the digital fields and innovation. The deep integration of technologies such as the new generation of 6G, IoT, digital communication and information and super computing power with urban development and strategy will jointly promote safer and more convenient and comfortable life for the citizens. The economic returns on China's investment in smart cities are astronomical. The GDP per capita of each new smart city is comparable to that of other mature big cities in the eastern coastal regions. China's technological advancement will soon overtake the US, whose trade wars and trade sanctions on China will have no significant effect.

8.2 New Smart City Social Structures

To live in a new smart city in China is to enjoy a harmonious, crime-free, green and pollution-free environment and respect for each other in the community. Everyone has time to enjoy walking , dancing, and strolling in the parks. Each city has parks and leisure facilities, as well as clubs for community residents to participate in cultural and social activities to promote the spirit of unity among the people. The roads are washed daily by automatic sprinkler systems. Each apartment in the condominium has a heat sensor connected to the fire station center via the IoT to inform firefighters of the actual location of the fire. The government has special fire fighting drones to fight fires on highrise buildings. The drones are remotely controlled. Here are some of the basic social structures for smart city living:

- Cultural activity

Each of the new smart cities has an open-air amphitheater that hosts concerts and art performances. There is also a large stadium that holds regular sports events with different seasons and competitions. The stadium is covered during the winter and rainy seasons to prevent bad weather from affecting the games and spectators. The stadium also has a banquet hall for marriage ceremonies. Various cultural organizations, such as calligraphy, Chinese chess, western chess, painting, pottery and other cultural activities, will hold regular gatherings for their members and the public. Primary and secondary schools and higher education institutions also organize cultural and sports activities for students, such as football, basketball, badminton, table tennis, billiards, tennis, hockey, swimming etc.

Both the amphitheater and the stadium are connected to a subway station. There are ample car park spaces for self-driving cars and buses. The residential area is served by trackless trains, which pick up passengers from various residential areas. Robotaxi are available 24/7 for different kinds of passengers at affordable prices with service from door to door and are very safe. All smart cities are connected with high speed rail with other cities throughout the country. Each smart city has a smart new airport to connect to every city in the country.

Every smart city has several libraries with books and magazines for citizens and students to borrow or read. The library also organizes public lectures and welcomes social experts to give speeches. Every smart city establishes a symphony orchestra supported by the city government and holds regular performances in the music hall. In this new smart city, the traditional culture is encouraged, and every festival will be celebrated. Each festival is organized by a group of voluntary associations for all citizens to celebrate enthusiastically.

A special day is chosen as the founding day of each new smart city and is marked by special events, celebrations, commemorations with mass participation by the residents. The city holds sports competitions with other smart city sports teams in football, basketball, table tennis, tennis, hockey , swimming. The city mayor encourages young people to participate in sports in their spare time instead of playing computer games. The city mayor encourages youths to join the scout association to develop their personal character to help others, to care for the community and the environment.

- Robot restaurant

Each neighborhood has several robot restaurants that operate 24/7 for breakfast, lunch, dinner and late night snacks, eliminating the need to hire human cooks and waiters. The daily menu can be found online. Robot restaurants across the city are connected to the IoT every day. Restaurants also have a variety of soups to suit individual preferences for a healthy diet. The restaurant has a meal delivery service for disabled persons, lonely people, single families, busy office workers, and housewives. The menu is designed for kids and dieters, with matching dishes. Daily health authority inspection of the food ingredients is done to ensure that they meet the health and safety requirements of the Health Department.

- Commerce and industry

The Central government has long term plans for every 5-year period to plan for economic growth and social development . Every smart city government has a responsibility to provide the right environment to support the industry and commerce to help China continuously move up the value chain and with continuous improvement. The city e-government establishes an R&D fund to help startups develop the latest IT applications, technological innovations, and products, aiming to strengthen scientific and technological capabilities to sustain the economy, develop intellectual property rights, and encourage enterprises to effectively utilize technology to increase productivity.

This is the digital age and the digital Yuan is being implemented by the government Central Bank for daily e-payment. The government invests heavily in increasing self-sufficiency in food production and renewable energy projects and improving infrastructures, promoting exports industries. The government holds regular meetings with the top management of industries to discuss the prevailing economic conditions around the world and seek advice where the government can assist to sustain their business.

The central government allocates sums of money to grant scholarships to promising students, to universities and enterprises to encourage research and development to create new industries. The government encourages private enterprises to spend on R&D to innovate and develop new products to keep abreast with time in the following fields : virtual reality, IoT, blockchain, AI, 3D technology, robotics (services, industry), new energy (lithium batteries, super capacitors, energy storage battery), new materials (graphene, carbon fiber), semiconductors, bio-based materials, biotechnology, life sciences, metallurgy, chemistry, powder science, medical devices, cloud computing.

8.3 SMART CITY ROBOTIC CONVENIENCE STORES

The unmanned 24/7 robotic convenience store will be widely accepted by the smart city communities. Unmanned 24/7 convenience stores incorporate image recognition, mobile payment, IoT and AI technologies. These unmanned stores are the main driving force for the current innovation and transformation of China's retail industry in neighborhood areas and in isolated places. These unmanned stores operate like the 7-Eleven convenience stores that are operating around the world to cater to customers to purchase a few items from the store. They do not operate like the supermarket where purchases by customers are in large quantities, and they are normally situated far from residential areas. China operates two types of unmanned stores, one for general provisions and the other for fruits. Unattended stores cover the whole country, including along the highway where rest stops are located.

Chapter 9 Asian Nations Association
9.1 Reason for Establishing Asian Nation Association

Since the end of WW II in1945, the world has undergone seismic changes in the world order. Hitler brought down the British Empire and helped build the USA as the superpower. Many colonies achieved independence from the colonial masters of Britain, France, Netherlands, and Spain. US stationed troops permanently on German and Japanese soil, making them an occupied semi-colony. The North Atlantic Treaty Organization (NATO) was created in 1949 by the United States, Canada, and several Western European nations primarily to provide collective security against the Soviet Union. The Warsaw Pact was established in 1955 by the Soviet Union as a military alliance against NATO. The Soviet Union was dissolved on 26 December 1991. However, the US and its allies did not dissolve NATO and still maintained it as a threat to Russia in a Cold War. The US also forms anti-China military alliances with Australia, New Zealand, India, and the UK in Asia to threaten China. The US did try hard to get ASEAN countries to join its anti-China military alliance but failed except the Philippines as the odd one joining US military alliance against China.

China has been facing the US military threat alone since 1949 when the Kuomintang government fled to Taiwan after it was defeated by Mao Zedong in a civil war. The US has maintained a fleet of navy ships , including an aircraft carrier, to patrol the South China Sea on the pretext of maintaining its right of freedom of navigation in international waters. The Western hegemonic power led by the US has been exerting enormous pressure and influence on individual Asian countries in trade, finance, politics, and international relationships for a long time. It is an appropriate time for Asian countries to unite and create an Asian Nation Association (ANA) to counteract against the West and US hegemonic power. Unity is strength. With unity amongst Asian countries, the US CIA will face difficulty to cause trouble to Asian countries individually.

With a population of 4.88 billion in 2023, Asia accounts for 60% of the world's population. Hence, Asian countries should have a voice on international platforms. At present, many Asian countries do not understand one another due to being colonized by foreign powers for many years. Also, post-independence countries are still immersed in friction amongst many political parties. Lack of education, skill and investment capital impede national economic progress. External power continues to interfere with domestic affairs of Asian countries, creating public discord , violent protests and toppling legitimate governments. Some Asian politicians are lured by CIA agents to accept bribes. Western colonial masters used to plunder the natural resources of the colony without compensation to local people. They have no interest in helping the ex-colonies to improve their economy and living standard. Some countries in the Middle East have been ravaged by bombings and killings by the US and the West. Hence, it is time for them to do some soul-searching on how to stop foreign invasion and regain their sovereign rights and security. Having oil is both a blessing and a curse. Oil can help a nation to proper but also invite hegemonic nations to invade the country thereby killing innocent people and destroying properties as has happened to Iraq, Syria and Afghanistan,

There are 48 countries in Asia and have diverse culture, history, religion, customs and different standards of education. It is not easy for these 48 countries to work together to achieve the desired goals. However, for a start, there needs to be a forum for every country to meet regularly and exchange ideas on how to implement projects for economic development. This should help each country to develop self-confidence and determination to fulfil their obligation in helping their country. China on BRI projects has built highways, railways, power stations, water supply, ports, schools and other infrastructure in Africa. These infrastructures are basic necessities in any country. Hence, China can help ANA member countries to build infrastructures to connect to the world.

ANA will have to establish a bank to offer loans to support any project approved by the ANA Council. China and Russia can offer scholarships to help good grade students who cannot afford to study in their local university. Each country should prepare a long term plan on how to improve its economy. Political stability is a prerequisite for social and economic progress. The priority for ANA is to prevent US and European hegemonic powers from infiltrating Asian internal affairs and the political arena.

9.2 CHINA AND ASIAN NATION ASSOCIATION RELATION

China has learned from the American proxy war in Ukraine, leading to immense human suffering from reckless killing of women and children. The war has caused destruction of Ukraine infrastructure and millions of refugees fleeing the country. The US wars in Iraq and Afghanistan also resulted in similar horrible fate to the people and countries. The US can create a proxy war with China by bribing the politicians of China's neighboring countries like the Philippines who allow the US to station troops on its soil. The Philippines needs to pay for the cost of military expenses and also paid $5.58 billion to purchase 20 F-16 fighter jets recently. The sole purpose of creating friction between China and its neighboring countries is for the US to sell weapons to them to fight China. In May 2023, Joe Biden gathered ASEAN leaders at the White House with a promise of $150 million in aid for their infrastructure, security, and other efforts aimed at countering the influence of rival China. The $159 million aid was a pittance equivalent to giving $0.234 per person to the population of 678 million people in ASEAN countries, compared with his offer of $5.9 billion aid to Ukraine to fight Russia in December 2024. In 2024 NATO led by the US attempted to recruit ASEAN members to form a NATO 2.0 in Asia.

Since 1946 the US CIA has been spreading propaganda of the "China threat" around the world, and also saying the Chinese government has no human rights. China has been reacting patiently all these years in rebutting the US propaganda through foreign affair official spokesman. Henceforth, the ANA secretary will demand the US to provide evidence about China committing genocide against the Uyghurs in Xinjiang, claimed by CIA propaganda. If no evidence is provided, the ANA will sanction the newsagent that spreads the disinformation. Such a response from the ANA carries more weight than China alone, as it represents 48 countries and 4.88 billion people.

The ANA can protest the atrocious bombing of Palestinian people in Gaza by Israel, assisted by the US and the West. With the approval of the General Assembly, the ANA can send a large peace keeping force to stop the war crime. The ANA can also sanction US oil export to ANA members if the US supports Israel in the war crimes against Palestinian. ANA has enormous power to counteract against the hegemonic power of the US and the West. In 2023, Asia's GDP is $40 trillion, surpassing America's GDP of $26 trillion. Asian countries control the raw materials supply chain. The US and the West need Asia.

The US already has a 5-Eye, Quad and AUKUS military alliance with the aim of containing China. On the issue of Taiwan and Hong Kong, the ANA could accept Taiwan and Hong Kong as representative members instead of official members. They can participate in various discussions at the General Assembly of the ANA, but they do not have voting rights. Chinese officials will have plenty of opportunities to meet Taiwan and Hong Kong representatives when they join the ANA. One day, China can propose a date for reunification with Taiwan at the General Assembly and seek support from the ANA members. Once the General Assembly votes in favor of reunification, China can set a final date for reunification. Once it has the support of ANA, China has no need to fear US opposition or send troops to defend Taiwan.

9.3 THE ASIAN NATION ASSOCIATION DEFENSE POLICY

The most important military provocation of the US in Asia is the stationing of military bases in Japan, South Korea, Qatar, Bahrain, Kuwait, Saudi Arabia, and the United Arab Emirates. These faraway military bases definitely are not meant to protect the US domestic security against external invaders. The military bases have been used to invade Iraq and Afghanistan. ANA General Assembly can pass a resolution to order the US to withdraw all their troops from member countries. There is no need for the US to station troops in Asian countries to defend their security, which can be done by ANA members themselves. From now onwards, there will be no subsidies from the ANA members to support the US military base. With unity, Asian countries can use domestically produced weapons for defense instead of using American weapons.

The ANA will not create a propaganda department like the US CIA to carry our smear campaign against any country. The ANA wants to maintain cordial and friendly relations with every country, including the US. The ANA members will henceforth develop their own defense industries and stop procurement of foreign weapons. The current foreign weapons will gradually phase out or be sold to other countries. The ANA will not enter any military alliance with other countries. ANA members carry out annual military exercises on land and the sea using weapons developed by the ANA members Other friendly countries are invited to participate. Member countries will not interfere with the domestic affairs of each country

War on terror is a smoke screen created by the CIA to infiltrate other countries' security organizations and political interference in the internal affairs of other countries. The US lures countries into joining the US anti-terrorism organization, where the US will provide other countries unverifiable "information" on potential terrorist attacks and offer assistance to stop the terrorist attack. Once a country joins the US anti-terrorism organization, the CIA will infiltrate this foreign country's foreign affairs department.

This US war on terror policy is a worldwide operation. Spreading rumors by the CIA about an impending terrorist attack is to sustain its ulterior motive of keeping alive American influence in the world. If no terrorist strike happens, the CIA claims it has succeeded in exposing the terrorist plan and preventing its attack. The prominent terrorist organization used by the CIA is ISIS. Turkish president Tayyip Erdoğan has accused the US of supporting ISIS in Syria, claiming Turkey has evidence of US support for ISIS. On 11 August 2016, Donald Trump accused Barack Obama and Hillary Clinton of being the "founders of ISIS." On 13 June 2024, Robert F. Kennedy Jr. the 2024 independent presidential candidate of the US presidential election, said during his foreign policy program speech: "We created ISIS."

The US has a persistent policy of wooing ASEAN countries to join it in a military alliance to contain China. The US regularly asks the British, German, and Australian navies to hold naval exercises in the South China Sea to provoke China.. When the ANA is formed, the Council can call a meeting to discuss the issue of US military incursion into Asian countries in the past and to remind the US not to repeat its hegemonic war in Asia. The US had created wars in Korea, Vietnam, Cambodia. Laos, Lebanon, Iraq , Afghanistan, Syria, Iran and Yemen, resulting in tens of millions of people perished. It is time for the US to leave Asia permanently for the sake of peace and security. ANA council will decide to set up a security force to maintain peace and security in Asia. The force has soldiers from member countries. Henceforth, Asia will have permanent peace without US military presence to cause trouble.

9.4 ASIAN NATION ASSOCIATION OBJECTIVES

A NA is not an anti-USA organization. Its core objective is to have a peaceful environment for economic and social development for all the Asian countries without external interference. The ANA will concentrate on helping members to develop their own economy, technical, financial and social advancement. Each country is encouraged to draw up a blueprint of projects on various developments in the next 5–10 years. Each country is to decide on the priorities of projects, technical skill, investment funds, and human and materials resources required. All the cross-trade transactions are done using each country's central bank digital currency without the use of SWIFT.

On international trade negotiation, the ANA will represent all members to negotiate import and export duties with other countries. This is similar to the EU, whose Commission negotiates trade relations with other countries. The ANA will not accept any country wanting to impose trade sanctions on any member country. The ANA will retaliate against any such sanction. The ANA will hold a meeting with the US and EU to discuss their sanctions on Russia, Iran, and North Korea to lift their sanctions entirely.

In order to help the ANA members to get to know each other better, there will be free visa travel to each other's country. People to people contact is encouraged by sports competition, arts and trade exhibitions, and cultural exchange. Hopefully, the ANA will help resolve the political conflict between Israel and Palestine. There will be no more armed conflict between them, as Palestine will be recognized as a country by the ANA Council. Every member state will have an embassy stationed in Palestine. There is a possibility that ANA can assist the Sunni and Shia to resolve their long-standing friction and help Palestine to become a country.

9.5 ANA MEDIA ORGANIZATION (ANAMO)

The aim of the establishment of the Asian Nation Association Media Organization (ANAMO) is to combat the biased reporting and fabrication of fake news by the European and American media. Most of the European and American media are no longer independent news reporting organizations. They have been used by the CIA to spread propaganda, fake news and misinformation against Asian countries. The CIA in 2024 has a budget of US$1.6 billion to spread anti-China propaganda and misinformation. The CIA has close working relations with US media which is privately owned and helps the CIA spread rumors, misinformation, and fake news worldwide. It also has similar working relations with western and Australian media.

The CIA spread unverifiable news about China committing Uyghurs genocide and hard labor in Xinjiang without evidence. When China invests in Africa, the CIA spreads news that African countries have now plunged into a debt trap. The American media did not publish CIA sponsoring Hong Kong young activists who demonstrated violently on the streets in 2019 with destruction of private and public properties, killing innocent pedestrians and causing chaos at the international airport. Nancy Pelosi said that the street riots and property destruction by the Hong Kong youth demonstrators was a beautiful sight to behold. She also met the activists in the congress to support their destructive behavior.

The CIA motive was to get the young activists who were paid by the CIA to topple the Hong Kong government and then destroy the Hong Kong economy. All the illegal activities committed by the CIA in

Hong Kong were never reported by US or Western media. Mao Zedong once said: "A spark can start a prairie fire." The Chinese government is constantly alert to the CIA creating a spark to cause a prairie fire in any region of the country by bribing terrorists or activists within the country or in neighboring countries.

The ANAMO has the power to demand US and Western reporters to show proof and evidence on any fake news, misinformation, or disinformation smearing any member of the ANA. If he refuses, he will be banned from operating in any of the ANA member countries. He will be sued and be prosecuted in court. Should he fail to comply with a court summons, he will be barred from entry into any Asian country. His bank account in Asia will be frozen. American social media sites such as Facebook, X (Twitter), Instagram and others will be required to self-censor fake news or statements that incite political unrest and conflict in the ANA countries. Failure to conduct self-censorship will result in the withdrawal of the license to operate in member states.

Chapter 10 Power Supply in China
10.1 Fossil Fuel for Power Supply

The power demand of a country depends on its economic strength, which in turn depends on its electricity production. China accounted for 19% of global GDP in 2023. China is the factory of the world and hence requires more energy for productions of goods for exports. As China does not produce enough oil to meet its requirements, it has to import oil. China is the world's largest crude oil importer, and in 2024, spending US$324.8 billion on oil imports. Rising oil prices have a profound effect on goods production and living costs. The US has used petrodollar to control the world's oil price, causing instability in the world financial market. China is concerned that escalating geopolitical tensions with the US could lead to the US blocking oil shipments from the Middle East through the Strait of Malacca, thereby cutting off China's oil supply.

To avoid disruptions in oil imports, China could make use of the annual oil-import budget of US$324.8 billion to invest in renewable energy projects, including hydropower, solar, wind, sea wave, nuclear and vegetable oil. The annual investment in renewable energy provides jobs to tens of thousands people. The investment money will remain in China and keep circulating in the economy.

Energy self-sufficiency is a national security priority. China is the world's largest battery manufacturer, and it requires a significant number of batteries to store excess power generated by its renewable energy plants. China's investment in renewable energy will have a synergistic effect on industrial and services growth, prompting more jobs and economic progress. China is the factory of the world and therefore consumes more power per capita than any other country. With self-sufficiency in energy, the Chinese government does not need to worry about oil supply and oil prices, which could cause inflation whenever the oil prices suddenly increase. When China's energy supply is stable, the industrial production costs will remain stable, making export prices remain competitive.

As of 2024, China's installed power generation capacity is about 3349 GW. In 2024 China's thermal power generation capacity reached 1,448 GW, accounting for 43% of the total power generation. As of now, coal-fired power generation is still China's main energy source. China imported a record of 542.7 million metric tons of coal in 2024. The price of thermal coal fluctuated widely, ranging from $440 to $48, over the past 10 years. China would spend tens of billions of USD annually on coal imports. According to a report, in 2024, China's renewable energy output is projected to account for 57% of its total energy requirement.

On September 26, 2021, a power outage suddenly hit the country. The power outage affected 20 provinces. Electricity has been cut in many places, causing many factories to open for four days and close for three days. Although China has a lot of power generation capacity, many provinces still face power shortages, affecting industrial operations and daily life. Presently more EV is sold in China and in the near future, EV will completely replace all internal combustion engine cars. As a result, China needs more electric power supply.

China wants to build 200 new smart cities to lift the remaining 600 million villagers out of poverty. Hence, more energy is required to build housing and infrastructures. China imported 2.17 million barrels per day of oil in 2024 with 1.57 million BPD from Saudi Arabia. According to a report, in 2023, China imported 22.7 billion cubic meters of Russian natural gas, with a significant portion delivered through the Power Natural Gas Siberia pipeline. The three primary countries exporting natural gas to China via pipelines are Turkmenistan, Russia, and Myanmar and 85% of China's LNG supply came from Australia, Indonesia, Malaysia, and Qatar. China has diversified its oil and gas suppliers. China RMB is gradually being used to pay for oil and gas imports.

10.2 ELECTRICITY PRODUCTION IN CHINA

China's total electricity production in 2024 reached approximately 10,000 TWh, compared with 8,632 TWh in 2021. China's renewable energy production in 2024 accounts for 57% of the total, and this component will be greatly increased in the future to reduce coal power generation and carbon dioxide emissions. The energy production of the US in 2024 was 4178 TWh for a population of 340 million compared with China's 1.4 billion. The US consumes more energy per capita than China. According to the experience of the US, wind and solar are the cheapest sources of energy production. China's electricity consumption includes power used to produce goods sold abroad. China is the world's largest manufacturing country. So China's power consumption also represents the power required for making goods for export.

According to report, in 2024, here are several countries in the world in terms of electricity consumption per capita (kWh per person) : US 12187, South Korea 11126, Australia 10046, Singapore 9576, Japan 7252, France 6069, Russia 6908, Germany 6227, China 6309, United Kingdom 3908, and India 1631. Although China is the second-largest economy in the world, the per capita energy consumption figure is way below that of the US. Energy consumption reflects the industrialization process of the whole country and the income level of individuals. China still has many poor people who cannot afford to consume more energy. And much of that electricity consumed is used to produce goods for export.

Due to the rapid increase in the use of electric vehicles, more cloud computing data centers and more high-speed rail operations in China, the demand for electricity will rise rapidly. Therefore, the government must immediately increase electricity generation to meet the rapid demand, otherwise more blackouts will occur. China spends billions of dollars a year on Belt and Road projects and imports oil and coal, but it has not invested enough in renewable energy projects. China has sufficient resources to implement full-scale renewable energy production, especially in wind and solar, to meet national demand.

10.3 POWER GENERATION CAPACITY OF CHINA

As of 2024, China's installed power generation capacity is about 3349 GW with about 57% coming from zero-emissions energy sources. This includes 27% from solar power, 16% from wind power, 13% from hydropower. China is stepping up efforts to build new renewable power generation capacity, adding 1,200 GW of solar and wind capacity by 2030. The power authority has warned that in some areas, including Inner Mongolia, Hunan, Hubei and Jiangxi, there would still be power shortages at peak times, despite the construction of new power plants and cross-regional transmission lines.

China has spent billions of dollars to rapidly build many high-speed rail lines, but has not built enough power generation to alleviate frequent power shortages, which directly affect people's lives and industrial production. Over time, power shortages become more acute as the wealth of the population increases. With AI, it is possible for China in the near future to produce enough power to satisfy national requirements as China has all the resources and technology. As of February 2025 China total installed capacity is a followed:

Share of capacity

Thermal Power	1448 GW	43%
Solar Power	926 GW	27%
Wind Power	530 GW	16%
Hydro power	437 GW	13%
Nuclear power	61 GW	2%

10.4 POWER SHORTAGE IN SOUTHERN CHINA

Relying on renewable energy sources for sustainable power supply is a challenge with unpredictable climate. According to a report, in the summer of 2021, Guangdong enterprises faced stringent "open four days, stop three days" electricity restrictions due to power supply shortage. Guangdong's electricity consumption increased by an average of 15.7% over the past two years, with the highest load demand exceeding 130GW, hitting a record high. Guangzhou, Dongguan, Zhuhai, Foshan, and other regions had started off-peak power consumption measures. In order to catch up with orders during peak season, some factories rent generator sets to generate electricity themselves as a back-up power supply. Meanwhile, Guangxi, Yunnan, Zhejiang, Hunan, Jiangxi and other regions also experienced power shortages. In 2021 rain came many days late resulting in the water supply in places such as Yunnan was about 11% less than expected, and hydropower production was reduced as a result. In order to achieve carbon neutrality by 2060, replacing coal power with renewable energy has become an irreversible trend. China needs to borrow a page from Denmark. As of 2023, wind power accounts for 57% of Denmark's total electricity consumption, compared with China's 16%.

China's per capita electricity consumption is only 6309 kWh per person per year, about half that of South Korea. China therefore needs to double down on building power plants. Experts suggest diversifying energy sources to address the risks of excessive reliance on hydropower in the South. China leads the world in both solar and wind power generation, but there is still much room for improvement, especially in northwest China. Although natural gas and nuclear power are not renewable energy sources, they are also low carbon emission scenarios to replace coal power. In 2025, the US estimated average cost of power generation type (US $/MWh):

Wind power onshore	25-45 $/MWh
Hydropower	30-80 $/MWh
Solar PV	30-50 $/MWh
Natural gas	40-60 $/MWh
Thermal power	65-150 $/MWh
Wind power offshore	70-150 $/MWh
Nuclear power	80-180 $/MWh

In 2023, China produced 86% of the world's solar panels and also had the largest wind turbine production capacity, accounting for 60% of the global total. Hence, China should invest more in renewable energy production using solar panels and wind turbines. China has the majority of the world's battery manufacturing capacity, with 77% of the global total in 2022. China's largest battery manufacturer, CATL alone, held around 35% of the global lithium-ion battery market in 2022 first quarter of 2022. Therefore, CATL batteries can be used to store power when there is excess power generated in any region.

Michael Loong

10.5 Power Grid System

China is a vast country, and every year, unpredictable and severe rainstorms, floods, and earthquakes occur in various regions, resulting in power outages. Prolonged power outages in big cities will seriously affect industrial operations, livelihood, traffic paralysis resulting in huge economic losses. The old manually operated power backup grid system cannot immediately cope with the need for emergency power delivery. When the demand for power exceeds the current supply, a new smart grid can automatically connect to the power backup immediately to start the power supply. Unstable energy demand requires a smart power grid to mitigate power interruption. Unstable wind and solar renewable energy generation, fluctuating consumer demand, power backup systems and a smart grid matching to achieve a balance between supply and demand becomes a complex project in a vast country like China. An AI solution is required.

The power grid is defined as the network connecting the power generation, transmission, and distribution units. Interconnecting power stations to each other is to allow the spare generating capacity as a rotating energy backup. Power is wasted if it is not used. The importance of the grid lies in the ability to instantly supply backup power to a region with power shortage. The distribution system is balanced whenever power supply and demand reach synchronous mode.

China's land mass is vast. When the sun has set in the eastern part of China, it is still shining in western part of China. Solar panels efficiency depends on the amount of sunlight . There are more sunlights in the western part of China, than in the eastern part. Renewable energy integration facilitates the transfer of electricity generated from sources

270

like wind and solar, which are often located far from consumption centers. Hence, ultra-high voltage (UHV) transmission lines are required to even out the variations in solar and wind generation by zipping electricity from a place that has favorable weather conditions to another area with less favorable weather conditions. Power generated must be used immediately, otherwise it is wasted. Interconnecting UHV grid lines can connect different regional or national grids, enabling more efficient energy distribution and balancing supply and demand.

UHV transmission lines are a critical component of modern electrical infrastructure, enabling efficient and reliable power delivery across vast distances. UHV AC transmission refers to AC transmission with a voltage level of 1000 kV and above. It has significant advantages such as large transmission capacity, long transmission distance, low line loss and space-saving. According to China Energy News, the combined length of the UHV transmission lines operating in China had reached 48,000 km (30,000 miles), more than enough to wrap around the Earth by the equator.

UHV transmission lines require special cables to carry the current. The cables core wires and insulations are specially made, heavy and expensive. Transmission towers must be robust to support the weight of the heavy cables and withstand the force of heavy wind and thunderstorms. Because wind and solar power is intermittent, UHV lines still rely heavily on coal or gas-fired power to ensure that their transmission is stable. It may be more cost-effective for cities and towns to generate power locally than to import them from long distances because it is expensive to build and maintain UHV grid lines. Some Chinese coastal provinces are planning to build nuclear and offshore wind power plants on a large scale so that they may not need to import electricity in the future, or at least reduce the amount of imported power.

To reduce transmission losses, smart cities can use the IoT to build a super smart grid construction and combine various renewable energy generation together with flexibility to adapt to changing demand. In 2023, the central government set up a huge battery power storage station in Guangzhou to provide emergency power needs. The Chinese company, CATL, the world's largest battery manufacturer which supplies batteries to many electric vehicle manufacturers, can supply the required quantity of batteries for power backup. IoT digital grid systems can improve power supply reliability and reduce power outages.

Special meters and sensors on transmission lines installed in homes and factories can continuously monitor demand and supply, while devices called synchronous phase-fluxes can measure the flow of electricity through the grid in real time, enabling operators to anticipate and avoid outages. IoT can collect all the data from power stations and consumers' electricity usage in real time. It can also automatically connect the power supply of each power station to maintain power distribution balance quickly.

10.6 GREENHOUSE EFFECTS FROM FOSSIL FUEL COMBUSTION

Greenhouse gases from burning of coal cause abnormal weather patterns in the global climate. Floods, droughts, snowstorms and extreme cold and heat waves have become frequent occurrences around the world. Record-breaking levels of climate change cause widespread damage to agricultural production, houses destroyed affecting daily life, leading to loss of animal and human lives. Coal burning also destroys ecosystems. Coal burning releases toxic chemicals that can cause arsenic and selenium poisoning, chronic obstructive pulmonary disease, esophageal and lung cancer. It also releases sulfur dioxide, which forms acid rain and damages ecosystems.

Fog and haze occur more frequently than before, and in severe cases toxic pollutants are trapped in the air in the atmosphere and are stopped from dispersing. Tiny particles can enter people's lungs and cause blood poisoning. This environmental pollution affects human health, leading to an astronomical medical and loss of labor costs. This toxic pollution also affects crops, livestock, poultry, aquatic products, and water sources. Increasing renewable energy generation projects is, therefore, an urgent task for the country's health and economic progress. Using cheap coal to generate electricity and then suffering from huge financial and human costs does not equate to a worthwhile human life.

Coal mining accidents are common in China, resulting in thousands of accidental deaths each year. As a result, the Chinese people pay a high price to use coal to produce energy, some of which is used for producing low value products for export earning meager profit margin.

Unfortunately, coal prices do not include the social costs of harmful side effects. China is the world's largest emitter of carbon dioxide, with about a third of its emissions being due to the production of goods for export. Countries that import Chinese products do not pay any pollution fees for coal power generation to compensate China for repairing pollution damages. This economic approach is not cost-effective and unfair to every citizen. Exporters should pay compensation for the pollution they create. In China, electricity prices are one of the lowest in the world.

History has shown that oil is one of the main causal factors of inflation. In the global oil market, OPEC members, oil speculators, and Western oil companies that control exploration, production, and sales have enormous power to manipulate the price of oil. When any war breaks out in the Middle East or elsewhere, the price of oil spikes and has a big ripple effect on the cost of living, even though there is no shortage of oil. The volatility of oil prices and the petrodollar exchange rate has caused chaos in the global financial market, resulting in slowing down of international trade and increasing global inflation.

In the past, every sudden increase in oil prices triggered an increase in fertilizer prices, which in turn led to a corresponding spike in global food prices and increased transportation costs, resulting in raising the cost of living to new heights. Inflation is a bogeyman for many common people, who fear to see their life savings for retirement visibly losing its purchasing power. Oil price volatility has significantly harmed the livelihoods of low-income people in China and elsewhere. The oil industry tycoons in the US and the West use oil as a tool to manipulate oil prices to fatten their pockets. US oil companies achieved record profit when the US and EU imposed sanctions on Russian oil exports.

The price of oil shot up to as high as US$115 per barrel from US$70 per barrel..

The US politicians have used petrodollar as a political tool in American foreign affairs for years. Every country must use USD to purchase oil and then settle the oil transaction using SWIFT, which is controlled by American banking institutions. Every country must pay foreign exchange fees to the American banking institutions, who earn enormous amounts of exchange fees for doing next to nothing.. The periodic rise of USD interest rates increases foreign exchange fees. In 2024, the national debt of the US will exceed 35 trillion dollars, and the inflation has reached a historical peak with oil prices in the US soaring to the peak. To contain inflation and the effect of oil price increase, the Chinese government must spare no effort in producing more solar panels and wind turbines to generate more power in order to reduce oil import as a long term solution and safeguard national security.

10.7 Hydropower Supply in China

In 2024, China's installed hydropower capacity is 437 GW, accounting for 13% of the total power generation of 3349 GW. Hydropower production was 1285 TWh, accounting for 12.9% of the total 10,000 TWh. In 2012, China's Three Gorges Dam became the world's largest hydroelectric power plant. The dam has a generating capacity of 22.5 GW. Since 2012, the Three Gorges Dam has been the world's largest power station by installed capacity. Its maximum annual power generation is close to 112 TWh. Baihetan Hydropower Station in Sichuan Province, the second largest in the world, is in operation in June 2021. Its total installed capacity is 16 GW, with 16 hydropower turbines of 1GW class, which is currently the largest single installed capacity of the hydropower station in the world. Its average annual power generation is 62.4 TWh. After the Baihetan project is put into operation, it will save about 19.8 million tons of standard coal and reduce carbon dioxide emissions by 51.6 million tons per year.

With a flood control capacity of 7.5 billion cubic meters, the Baihetan Dam will serve as a flood barrier for the middle and lower reaches of the Yangtze River. During the dry season, it will also increase downstream flow and improve navigation conditions. The Baihetan Dam is embedded with thousands of sensors that can measure important information such as temperature, deformation, and stress and can be transmitted in real time to the hydropower station's management. Such feedback information allows accurate intelligent control and real-time adjustment of various parameters of the project, making the Baihetan Dam the most intelligent dam in the world.

China has approved a $137 billion project to build the next most powerful hydro dam in the world. It will be located in the lower reaches of the Yarlung Zangbo River in Tibet, which measures 1,125 km in length. According to a 2020 estimate by Power Construction Corp of China, the new hydro dam could produce 300 TWh electricity annually. Should construction go ahead, the dam is set to take 35 years to complete. The Yarlung Zangbo Hydroelectric project is part of China's 14th Five-Year Plan, which aims to reduce pollution and increase renewable energy. The dam is expected to generate three times more power than the Three Gorges Dam.

10.8 CHINA'S WIND ENERGY SUPPLY

In the first half of 2024, China's wind power capacity had increased by 19.9% year-on-year to 39.1 GW. China is on track to reach 1,500GW of installed wind and solar capacity by the end of 2024. China is the world's largest manufacturer of wind turbines. The government plans to increase the supply of wind power exponentially to reduce the supply of coal-fired power. Wind power is used as the main supply for the national grid and backup power. The installation of offshore coastal wind turbines takes advantage of the naturally relatively large temperature difference between the ocean and land to cause high speed winds and volumes.

Wind speed is relatively high in desert areas, so it is suitable for large wind farms to install power supplies. A fifth of China's land is covered by desert, which can generate a lot of power. Using this power source and with thousands of canals, river water is transported to desert areas across the country, creating reservoirs, lakes, and marshes. At the same time, the whole desert starts to plant trees, grass, vegetation, restore desert ecology, to stimulate economic activities, including oases and wildlife survival. The severe sandstorms that Beijing and other northern cities endured for years will be eradicated forever. Blue skies and fresh air will provide a better life and a healthier environment.

China's total installed wind power capacity in 2024 reaches 530 GW and wind power generation would reach 989 TWh accounting for 9.9% of total power generation. Wind power will remain the third-largest source of electricity in China in 2024, accounting for 16% of total power generation capacity. With the Chinese government laying out a roadmap to reach 1,500 GW of total installed wind and solar capacity before 2030, China has identified wind power as a key component of the country's economic growth. China has abundant wind power to allow China to build up to 2.38TW of wind power capacity in the future, enough to eliminate the need for coal power generation. China makes wind turbines, so it does not need to rely on foreign suppliers. China also has the necessary experience in installing wind farms on land and offshore. There is no reason that China does not want to increase wind and solar power generation now, especially since the current Chinese economy is sluggish with high unemployment.

10.9 FOUR MAJOR WIND FARMS IN CHINA

Besides hydropower, wind power is the renewable energy with the most mature development technology and the lowest development cost. It is the renewable energy with the most large-scale development value in the future. As the manufacturing of wind power equipment has improved, the cost of wind power construction has also decreased, and the economies of large-scale development have also increased. Therefore, the development of wind power is an inevitable choice for China's renewable energy to replace coal power. There are four major wind farms in China: Jiuquan 10GW wind farm in Gansu Province, Dabancheng Wind Farm in Xinjiang, Huiteng Xile Wind Farm in Inner Mongolia and Rudong County Offshore Wind Farm in Jiangsu.

- Jiuquan Wind Power Base

Gansu Jiuquan Wind Power Base is a group of large-scale wind farms built in western Gansu located in the desert area near Jiuquan City, Guazhou County, with relatively abundant wind power and an installed capacity of 20GW in 2020. The project is the world's largest wind farm, costing 120 billion yuan. Jiuquan Wind Farm is known as the Three Gorges of wind power. It is reported that after the completion of the wind power base, the project reduces emissions of sulfur dioxide by 8,071 tons and 715000 tons of carbon dioxide with annual saving of 250,000 tons of coal.

- Xinjiang Dabancheng Wind Farm

Xinjiang is bestowed with excellent wind energy resources. According to reports, in view of the valley between the central and eastern Tianshan Mountains, the area where wind turbines are to be installed is more than 1,000 square kilometers. At the same time, the wind speeds distribution here is relatively stable. Dabancheng Wind Power Farm has an annual wind energy reserve of 25 TWh, and the total electricity available is 7.5 TWh, with an installed capacity of 2.5 GW. In 1988, Xinjiang received a grant from the Danish government to complete the first phase of Dabancheng Wind Farm. It is the earliest wind power farm in the autonomous region and the earliest experimental site for developing wind energy in China. The winter period is the peak of wind power operation in Dabancheng because the winter brings huge wind speeds, promoting higher economic efficiency of power generation output.

• Huiteng Xile Wind Farm in Inner Mongolia

Huiteng Xile Wind Farm is located in the Inner Mongolia Plateau, with high sea level and abundant wind resources. Featuring strong stability of wind power density, good sustainability and high quality of wind energy, it is the most ideal place for the construction of wind farms. Since the construction of the wind farm began in 1996, 94 wind farms have been installed. By the end of 2009, the total installed capacity of Huiteng Xile Wind Farm reached 1.085GW.

• Rudong County Offshore Wind Farm in Jiangsu

The 2.32 billion Yuan Rudong offshore wind project covers an area of about 130 square kilometers. It can contribute 375 GWh of clean energy electricity to the grid every year, becoming the largest offshore wind farm in the country, with a total installed capacity of 182 MW. 80 wind turbines were installed. It will be the first batch of the largest offshore wind power cluster to be built and put into operation during the "14th Five-Year Plan" period. There are 150 huge wind turbines spinning around, which can meet the electricity needs of 1 million homes for an entire year. The Rudong Wind Farm will reduce carbon dioxide emissions by 1.46 million tons per year. The wind power facility will help ease power shortages in East China's Jiangsu province, the country's economic engine and manufacturing hub.

10.10 SOLAR POWER GENERATION IN CHINA

China is the world's largest producer of solar energy. In 2024, China's cumulative installed capacity of renewable energy power generation reached 1836 GW, accounting for 57% of the total installed power capacity. China added 163 GW of solar capacity in the first nine months of 2024, which is equal to the combined solar capacity of Germany, Spain, Italy, and France. Solar power generation is an important energy source for the future world, and solar energy is inexhaustible and completely environmentally friendly. Solar thermal power systems have a solar collector, which has two main components: a reflector (mirror) that captures and focuses sunlight onto a receiver. In most types of systems, the heat transfer fluid is heated and circulated in the receiver to produce steam. The steam is converted into mechanical energy in a turbine, which drives a generator to generate electricity.

Solar thermal power systems have tracking systems where solar energy is focused on the receiver throughout the day when the sun's position in the sky changes. Solar thermal power plants usually have a large field or array of collectors that provide heat to turbines and generators. A solar thermal power system can also have a thermal energy storage system that allows the solar collector system to heat the energy storage system during the day and the heat from the storage system to be used to generate electricity at night or on cloudy days.

There are three types of concentrated solar thermal power systems: Linear concentrating system, including parabolic groove, solar power tower and solar disk/engine system:

- Linear concentrating systems use long, rectangular, curved (U-shaped) mirrors to collect solar energy. The reflector focuses sunlight onto a receiver (tube) that is the same length as the reflector. The concentrated sunlight heats the fluid flowing through the pipe. The fluid is sent to a heat exchanger, which boils water in a conventional steam turbine generator to produce electricity. There are two main types of linear concentrating systems: parabolic trough systems, where the receiving tube is located on the focus line of each parabolic mirror. Linear Fresnel reflection systems, in which a receiving tube sits above multiple mirrors, allow the mirrors to have greater fluidity when tracking the sun. Linear centralized collector power plants have a large number of collectors, which are usually arranged in a north-south direction to maximize the collection of solar energy. This configuration allows the mirror to track the sun from east to west during the day and continuously focus sunlight onto the receiving tube.

A parabolic trough collector has a long parabolic reflector that focuses the sun's rays on a receiving tube located at the focal point of the parabola. When the sun moves from east to west during the day, the collector tilts with the sun to keep the sunlight focused on the receiver. Due to its parabolic shape, a slot can focus sunlight from 30 to 100 times the normal intensity (concentration ratio) on the receiving pipe, located at the focus line of the slot, achieving an operating temperature above 400C°.

- The solar disk/engine system uses a mirror disk, similar to a very large satellite disk. To keep costs down, mirror plates usually consist of many smaller flat mirrors that form the shape of a plate. The disk-like surface directs and focuses sunlight to a thermal receiver, which absorbs and collects heat and transfers it to the engine generator. The most common type of heat engine used in the disk/engine system is the Stirling engine. The system uses fluid heated by the receiver to move the piston and generate mechanical power which drives a generator or alternator to generate electricity.

- The solar tower system uses a large flat array of sun-tracking mirrors, called heliostats, to reflect and focus sunlight onto a receiver at the top of the tower. Sunlight can be concentrated up to 1500 times. Some power towers use water as a heat transfer fluid. The thermal energy storage capability enables the system to generate electricity in cloudy weather or at night. Qinghai's Gonghe 50 MW solar thermal power generation project is located in Gonghe Hainan ecological solar power park. In 2021, the all-day online power reached 539 MWh. The project uses a tower type concentrated solar thermal power generation direct air cooling unit, equipped with a solar heat collection tower. The plant can produce 157 GWh of electricity annually during its operational period. The solar thermal power generation project integrates power generation and energy storage.

Built in the Gobi Desert in Gansu province, a solar power station project consists of 12,000 mirrors in concentric circles around a 260-meter-high column heat collector tower. The mirrors act as receivers of solar energy, reflecting sunlight to the heat absorber at the top of the tower, allowing the molten salt in the absorber to absorb heat, raising its temperature to 550 degrees Celsius. Molten salts transfer the heat to the steam generation system. There, they exchange heat with water and hot gas to produce high-temperature, high-pressure steam, which then drives the turbine generator set. The thermal salt station can generate electricity continuously for 24 hours, generating 2.4 GWh per day.

It is reported that China will implement a large-scale solar photovoltaic power plant in the Kubuqi Desert renewable energy base in Inner Mongolia. The first project could start construction in 2025 and is expected to be commercially operational in 2027, providing 16 GW of electricity to more than 1 million homes when fully operational. The Kubuqi project is vast, with some 225 bases in the deserts of western and northern China. By 2024, the total power generation capacity of all renewable energy projects will reach 1456 GW, of which 64% will be solar and the other 36% will be wind. No country other than China currently has such a large amount of clean energy generation capacity. China's renewable energy is expected to reach 1500 GW before 2025. Solar power is the second-largest power supply in China.

10.11 Space Solar Power Station

Solar energy is inexhaustible. China plans to have space power stations where a large array of solar panels will be set up in orbit to collect sunlight to produce electricity which is sent back to a receiving station on Earth by laser beam. The power station is able to supply electricity 24 hours daily. The space power station weighs about 1,000 tons, compared to 400 tons of the International Space Station. China uses 3D printing technology to build the power station. According to a report, on May 5, 2020, China used a Long March 5 rocket to carry a 3D printing machine to the space station, starting to make fiber reinforced composite materials to make power plant materials, and then continued to ship some raw materials to build the final power station. The projects begin to build small and medium-sized solar power stations for experiments. About a few years later, a megawatt-class experimental space solar power plant would be built and launched into a 36,000 km high geostationary orbit. Some years later, China began to build a GW size commercial space solar power station operation. According to a report, space-based solar panels can generate 2,000 GW of power constantly, and hence China does not need to import fossil fuels for power generation.

10.12 Nuclear Power Generation

Although nuclear power is a clean source of energy, the cost of nuclear reactor meltdowns is enormous. There have been three serious accidents with nuclear power plants: Three Mile Island in the US in 1979 and Chernobyl in Ukraine in 1986. Radiation from these accidents remains in the affected areas for years. Japan's Fukushima nuclear plant was damaged by a tsunami in 2011 and permanently disabled. Japan released large amounts of treated radioactive water from the Fukushima nuclear power plant into the Pacific Ocean in 2023, which was opposed by countries around the world, with only the US endorsing the Japanese measure. China in 2024 has nuclear power plants generating 61 GW, with 22 under construction to generate 24 GW. More than 70 are planned for construction in the future, generating 88 GW of electricity. China aims to achieve a zero carbon footprint by 2060.

In 2025, China's Zhangzhou nuclear power plant completes a significant milestone with the hot functional test of its second unit, setting the stage for fuel loading. The Hualong-1 reactor enhances China's global competitiveness in third-generation nuclear technology, serving as a model for sustainable energy solutions. With plans for six reactors, the plant will generate over 10 TWh of clean electricity per reactor each year, supporting China's long-term energy strategy. The plant's strategic objective is to establish a large-scale energy hub, consisting of six nuclear power units, each with a capacity in the million-kilowatt (GW) range. The Zhangzhou nuclear power plant is not only a technological marvel but also a cornerstone of China's economic and environmental strategy. The plant's full operation will yield an installed capacity of approximately 7.2GW. This output can power the annual electricity needs of one million people in moderately developed countries, highlighting the plant's potential impact on energy security.

10.13 ELECTRIC VEHICLES (EV) IN CHINA

In 2024, China will produce more than 10 million EV, the highest in the world. It is also the first time China exports more cars than Toyota. China's new energy vehicle ownership accounts for about 58% of the global total. From January to September 2024, a total of 11.5 million EV sold worldwide, with the Chinese market contributing 7.2 million of them. When the EU imposes a tariff of up to 38.5% in Oct. 2024 on Chinese EV imports, China lodges a protest. Later, the EU relented on its tariff imposition of Chinese EV, but introduced a new policy of asking China to produce EV in EU by transferring technology property rights to local companies. Chinese EV and battery investment projects in the EU are then suspended. China also faces a tariff of 100% imposed on Chinese EV by the US. With the headwind in the US and EU EV market, China EV exports expand to countries in the Middle East, Central Asia, ASEAN, Africa, and South America.

Many Chinese EV manufacturers announce plans to invest heavily to build new factories in Thailand, Malaysia, and Brazil. They have opened showrooms in Zambia, Kenya, and South Africa. In Thailand, companies such as Great Wall and BYD are leading the way to open the market there. BYD opened a production facility in Thailand in July 2024. In addition to Thailand, BYD has also captured a large market share in Singapore and Malaysia. According to government statistics, the EV behemoth ranked as Singapore's second-most popular car brand by sales in the first half of 2024. BYD ranked among the top 10 car brands in Malaysia. BYD plans to launch a partnership with Uber aiming to bring 100,000 BYD EVs to Uber drivers globally. In addition, BYD is planning a new auto factory in Brazil to come online in 2025. Both BYD and Great Wall have local R&D, production and sales centers in Brazil. In March 2024, China partnered with Ethiopia plans to introduce nearly half a million EV in Ethiopia over the next decade.

China also dominates the plug-in electric bus and light commercial vehicle global market, reaching over 500,000 buses (98% of global stock) and 247,500 electric commercial vehicles (65% of global stock) in 2019. 447,000 commercial EV sales were recorded in 2023. BYD has overtaken Tesla as the world's largest maker of EV . About 650 electric buses currently on the road in the US, nearly 400 come from BYD, and all of them are manufactured by BYD factory in Lancaster, California. In 2023, BYD revenue reached 483 billion yuan, producing more than 3 million vehicles. BYD has an advantage over legacy automakers with its vertically integrated supply chain. It makes almost all components of its cars in-house rather than farming them out to suppliers. According to Bloomberg, there were 500 Chinese EV car manufacturers in China in 2019. After fierce competition, only 100 manufacturers remained by 2023

CATL is one of the first internationally competitive power battery manufacturers in China. In August 2023, with a revenue of 48.85 billion US dollars, it was listed on the 2023 Fortune Global 500 list. CATL battery is currently used by EV manufacturers in the international market, including BMW, Daimler, Hyundai, Honda, Li Automotive, NIO, PSA, Tesla, Toyota, Volkswagen, Volvo and Xopeng. Nearly half of Tesla's vehicles use LFP batteries supplied by CATL. By 2021, it will have been the world's largest manufacturer of EV batteries for five consecutive years. In 2021, CATL's share of the global automotive lithium-ion battery market was 32.6%. This figure is expected to grow to 34-37% in 2024.

Chapter 11. China Water Resources
11.1 Current Status of Water Supplies in China

China has suffered decades of floods in the south and droughts in the north. These natural disasters destroy crops and livestock. The farmers' lives were ruined. China spends a lot of money every year importing food to feed its population and livestock. In addition, industrial CO_2 emissions and the widespread use of agricultural pesticides have polluted water sources, and acid rain falling from the sky has polluted rivers and farmland. Polluted rivers and oceans kill fish species and microorganisms. This situation cannot go on forever and the Chinese people will continue to suffer. China forbids fishermen from going out to sea for several months each year.

Water is the source of life. China has 20% of the world's population but uses less than 6% of the world's water resources for food production. China's total freshwater resources are 2.8 trillion cubic meters, accounting for 6 percent of global water resources. China's per capita water resources are only 2,300 cubic meters, only a quarter of the world average. China has one of the poorest per capita water resources in the world. China depends on rivers, lakes, and underground rivers for clean water supply. With China's urbanization, economic, industrial and social development, population and foreign trade growth, the demand for water resources is increasing, but the natural clean water resources are in short supply. Yet, clean water is increasingly being polluted. Such a dire situation is not conducive to nation building and economic advancement.

Due to its large and diverse geography, China has a wide range of topographical and climatic zones that result in an uneven distribution of water resources. The north is lacking in water resources, while the south is relatively abundant in water resources, and sometimes too much rain water leading to floods. The biggest flood occurred in China in 1931, when several major rivers in China, such as the Yangtze River, the Pearl River, the Yellow River, and the Huaihe River, suffered from severe overflowing. The scope of the flooding extended to the Pearl River Basin in the south, beyond the Great Wall Pass in the north, from northern Jiangsu in the east to the Sichuan Basin in the west.

The flood was widely considered to be the deadliest natural disaster of the 20th century, with an estimated 4 million deaths. The Yellow River's serious silting problem has caused flooding throughout history due to rising riverbeds. Recently the river excavation project has removed 300,000 tons of silt from the Yellow River, resulting in the riverbed dropping 5.1 meters. Riverbed excavation will continue to remove sands and sediments from the upper reaches of Loess Plateau. Meantime, more trees are planted on the Loess Plateau upper reaches to prevent soil erosion.

China has invested significant resources and hundreds of billions of dollars in developing highways, high-speed rail, and modern airports. However, insufficient effort and investment have been directed towards securing adequate clean water to sustain growing crops, operate industries, raise livestock, and prevent desertification. This shortfall necessitates China's expenditure of hundreds of billions of dollars on importing large quantities of grain and meat from abroad. In 2023 China imports 100 million tons of soybeans from the US and Mexico, valued at US$58 billion and US$ 4 billion spent on import of 12 million tons of wheat. One day, the US government might impose sanctions on

China, stop the export of grain and meat to China, and pressure other countries to do likewise. In order to avoid any food crisis in the future, China must immediately build hundreds of high-tech AI-powered agricultural farms to provide the population with food self-sufficiency.

The Chinese people have depended on the Yangtze River and the Yellow River to sustain their lives for thousands of years. However, both rivers have been severely polluted for a long time, and water pollution continues to occur despite the government's active anti-pollution efforts. China must solve the pollution problems urgently to achieve an unlimited supply of clean water soon. China has sufficient capital, skills, technology and human resources to accomplish this task.

11.2 IMPORTANCE OF YELLOW RIVER

The Yellow River, also known as the Mother River of China, originates in the Bayankala Mountains of Qinghai Province. It flows through nine provinces and regions of Qinghai, Sichuan, Gansu, Ningxia, Inner Mongolia, Shaanxi, Shanxi, Henan and Shandong, and finally flows into the Bohai Sea in Shandong Province. The Yellow River is 5,464 kilometers long. Its drainage area is about 750,000 square kilometers. The main tributaries of the Yellow River are Huangshui, Baihe, Heihe, Taohe, Qingshui River, Dahei River, Groye River, Fenhe, Wuding River, Jingwei River, Luohe, Qin River, Jinti River, and Dawen River. The 14 tributaries, spread across nine provinces from Qinghai in the upper reaches to Shandong in the lower reaches, provide drinking water and irrigation to 140 million people. The Yellow River is one of the longest rivers in the world and the second-largest river in China.

The Yellow River drains an average of about 53.5 billion cubic meters of water into the sea every year, with a flow rate of about 1,770 cubic meters per second. Much of the Yellow River is heavily polluted and unsafe for agriculture and industry. It is called the Yellow River because it carries more than 1.6 billion tons of silts every year, with the silts deposited in the lower reaches of the river, thereby raising the riverbed and creating overflow and flooding. In the more than 2,000 years of recorded history, the lower reaches of the Yellow River had broken their banks more than 1,500 times and changed their course 26 times. Every time the levees burst, they caused heavy damage.

In 1933, 54 downstream levees broke, affecting an area of more than 1 million square kilometers and more than 3.6 million people. After the flood, the disaster area was ravaged by diseases, plague, and death. Millions of people were displaced and became homeless. Most of the sediment in the Yellow River came from the Loess Plateau. The Loess Plateau is one of the largest and thickest loess plateaus in the world. Its 635,000 km2 area corresponds to around 6.6% of the land area of China. Around 108 million people inhabit the Loess Plateau. Hence, it is a monumental task to tame the silting problem caused by the huge area of the Loess Plateau.

The government had implemented a system to combat the soil erosion from the very tops of the hills to the bottom of the ravines of the Loess Plateau. Over 100 species of woody trees and more than 30 types of grass were planted, and a vegetation coverage of over 75 percent of land had achieved a remarkable sediment trapping rate of 98 percent. The Yellow River Basin is rich in fish resources, with more than 160 species of native fish of 92 genera, of which 19 species are unique in the world. However, due to habitat loss, pollution, and overfishing, many native species have declined or disappeared altogether. China's total freshwater resources are 2.8 trillion cubic meters, accounting for 6 percent of global water resources. In the long history of China, the Yellow River was regarded as both a blessing and a curse, and was called "the pride and the sorrow of China."

11.3 Plight of the Yellow River Water Supply

According to reports, the annual average natural runoff of the Yellow River is 53.5 billion cubic meters. The per capita water volume in the Yellow River basin is 593 cubic meters, which is 25% of the national per capita water volume. The Yellow River has a large sediment content, with an average annual sediment of 1.6 billion tons. The Yellow River, with its limited water source accounting for 2% of the country's river runoff, undertakes the task of supplying water to 15% of the country's arable land, 12% of the population and more than 50 large and medium-sized cities in the basin and downstream Yellow River irrigation areas, as well as long-distance water supply to some areas outside the basin. The water resources of the Yellow River not only supply water for the national economy inside and outside the basin, but also leave a certain amount of water to maintain the ecological environment of the basin.

It is reported that the river runoff consumption is 30 billion cubic meters, and other factors directly or indirectly consume the river runoff, estimated at 5-8 billion cubic meters. Therefore, the actual consumption of surface water in the Yellow River has reached 35-38 billion cubic meters, accounting for more than 60% of the annual average natural river runoff of the whole river. It is generally believed that water usage should not exceed 40% of river runoff, however, the utilization of water resources in the Yellow River has exceeded this limit. This situation increases risks to ecological health and the ecosystem of the Yellow River.

With climate change, the rainfall runoff in the Yellow River basin is getting less, and the water consumption of the Yellow River regions by various consumers is increasing. Such a situation has caused the water environment to deteriorate in the lower reaches of the Yellow River and its tributaries. The river channel is shrinking, and the sediment deposition in the river channel is increasing, resulting in the imbalance of the supply and demand of the Yellow River water resources. Water shortage had caused supply interruption of the lower reaches of the Yellow River in 1979 and 1997. The main tributaries, including the Wei River, Fenhe River, Yiluo River, Qinhe River, and Dawen River, had been cut off. Urban water shortage was becoming more serious in cities such as Hohhot, Xi 'an, Taiyuan, Xianyang, and Tongchuan. Urban water shortage had caused serious impact on people's life, industrial and agricultural production, and groundwater overexploitation had brought serious ecological and environmental problems.

With the rapid economic development of the Yellow River basin, it is expected that by 2030, the increases in population, urbanization rate , industrial output and food demand will exacerbate the water resource shortage problems. Six provinces and autonomous regions in the upper and middle reaches of the Yellow River, including Qinghai, Gansu, Ningxia, Inner Mongolia, Shaanxi, and Shanxi, are predicted to have a water shortage of 11 billion cubic meters and 16 billion cubic meters respectively in 2030 and 2050 under normal water conditions. With the future diversion of river water from estuaries to the interior and the construction of thousands of new reservoirs, the future water shortage problem could be resolved. With the establishment of AI-powered agriculture factories, water consumption and pollution will also be reduced and controlled.

Rice farming requires a lot of water to grow. Rice farming consumes 70 percent of China's agricultural water resources. Too much water overwhelms the rice, and too little can stunt growth. The evaporation of water also contributes to an increase in irrigation water consumption. It is reported that an average of 15 tons of water is needed to produce 1 kilogram of beef. Increased demand for beef has led to overgrazing. Grasslands become semideserts due to a lack of water, and ecosystems fail to function, leaving the land ultimately drier. It is estimated that farmers currently use 1,000 tons of water to produce a ton of wheat worth $285 and farmers don't have to pay for water. Traditional farming uses a lot of water and also wastes a lot. China will use AI robotic farms to produce crops, thereby saving a lot of water.

11.4 IMPORTANCE OF THE YANGTZE RIVER

The Yangtze River originates from the Tanggula Mountains (Tibetan Plateau) and runs for 6,300 kilometers to the East China Sea. In terms of water flow, it is the third-largest river in the world. Its river basin covers one-fifth of China's land area and is home to nearly one-third of the country's population. The Yangtze River has a lot of water. In the upper reaches, with the average discharge exceeding 1,980 cubic meters per second, which is more than the Yellow River. The Yangtze River flows through 11 provinces of Qinghai, Sichuan, Xizang, Yunnan, Chongqing, Hubei, Hunan, Jiangxi, Anhui, Jiangsu, and Shanghai.

Severe pollution of the Yangtze River, mainly caused by industrial wastewater discharge, agricultural fertilizers, sedimentation, ship waste and acid rain, also exacerbates the impact of seasonal flooding. It is reported that about 30 billion tons of waste water flows into the Yangtze River every year. A large amount of rainfall flows into the Yangtze River basin. Floods from the Yangtze River have kept the region's soil fertile. Many of China's cities are located along the Yangtze River and are home to more than 400 million people. The prosperous Yangtze River Delta generates 20% of China's GDP.

According to statistics, the annual runoff of the Yangtze River is as high as more than 960 billion cubic meters, while the total flow of the Yellow River is about 53.5 billion cubic meters. The water of these two big Chinese rivers flows into the sea every year and disappears whilst the country's more than 700 cities, most of the northern cities, face the problem of insufficient water supply. Moreover, China is not so rich in water resources. Many people in remote places need to walk hundreds of meters to fetch water from the wells. The bigger water problem is that 90% of China's urban groundwater is also contaminated.

11.5 Severe Drought in 2023

In 2023, the following areas of China were severely affected by climate change:

- Sichuan, Chongqing, Hubei, Hunan, Anhui, Jiangxi and other regions with
- 32.99 million mu, 2.46 million people, 350,000 head of large livestock affected by drought.
- The water surface area of Dongting Lake and Poyang Lake had shrunk by three quarters. Poyang Lake, China's largest freshwater lake, had recorded the lowest water level in 71 years, and the lake bed had turned into grassland. The water level in Ezhou section reached the lowest level since hydrological records began in 1865
- Chongqing's highest temperature reached an unprecedented 45 degree C.
- Power plants were affected by the lack of water. The Sichuan government
- extended power rationing to industries from six days to 11 days.
- Across the Yangtze River basin, 66 rivers had been cut off and 25
- reservoirs had dried up.
- A total of 6.58 million people in 17 cities and 85 counties in Hubei had been
- hit by drought, with 1.02 million mu of crop loss and 5.8 billion yuan in economic losses.

11.6 WATER SCARCITY CAUSES SERIOUS SOCIAL PROBLEMS

It is reported that the lack of water in China causes a national economic loss of more than 200 billion Yuan every year. For many years, severe drought in the north has affected large areas of farmland and the lives of millions of livestock and people. In 2009, five western provinces experienced the worst drought, affecting more than 61 million people. Drought has caused many important lakes and wetlands to become dry land, affecting ecosystems and groundwater sources. Many rivers flowing through cities and villages in many provinces had been intercepted by people to irrigate farmlands, to support industries and provide water for urban consumption. As a result, the amount of water discharged into the downstream lakes gradually decreased and the lakes dried up.

The serious water shortage has constrained the modernization of China's cities and towns, the growth of GDP and the improvement of people's living standards. Prices of daily consumer products rose, causing inflation. Water scarcity creates a cascade of serious social, national security, health and economic problems. Water is the source of life. Urban water demand continues to grow, but the amount of urban water resources is limited. The local water resources in most cities had approached or reached the limit of development and utilization, and the groundwater in some cities had been in a state of over exploitation.

When groundwater extraction exceeds replenishment, the quality and quantity of water resources will be out of balance, and a series of environmental geological problems will surface. Excessive exploitation of groundwater causes the groundwater level to decline year by year that could result in ground subsidence, collapsing, cracking and other problems. As a result, the government spends a lot of money every year to repair the substantial damages. All this money spent each year brings no economic benefit to the country. Why not budget that amount of money to solve the water shortage problem forever? Spending money first on preventing water problems is better than spending money on disaster relief. Disease prevention is always better than cure.

11.7 COSTS OF WATER POLLUTION

Water pollution in China is man-made and the costs of pollution are astronomical. China's water pollution has caused serious economic losses, as the country feeds more than 20% of the world's population with less than 6% of the world's arable land, while its land area continues to shrink due to environmental problems. According to statistics, 12 million tons of grain in China are contaminated by soil heavy metals every year, mainly from mining and smelting processes, industrial waste, tannery wastewater, textile mill wastewater, resulting in losses of up to 20 billion Yuan per year.

At present, the problem of soil pollution in China not only affects traditional agricultural fields, it also involves harm to public health caused by food poisoning related to soil pollution. Soil pollution refers to the phenomenon of heavy metals, pesticides, detergents, antibiotics and persistent toxic organic substances entering the soil and exceeding the self-purification capacity of the soil. Finally, it results in changes in the physical, chemical and biological properties of the soil, reducing the yield and quality of crops, and endangering human health.

Soil pollution has the characteristics of being latent and persistent, as the soil pollution can last for centuries. Soil pollution from contaminated water affects crops that could enter the human body, thereby affecting the human cells. The varied sources of soil pollution also bring difficulties to the prevention of water pollution problems. The free interaction between the soil and natural elements, such as the atmosphere and water, promotes various scales of pollution, thereby increasing the difficulty of treatment. For example, the soil pollution caused by landfill can penetrate groundwater through rainwater, and the polluted groundwater enters the rivers and causes soil pollution in irrigation areas. In 2012, after a mining company contaminated two rivers in southern China with cancer-causing cadmium, officials warned about 3.7 million residents of Liuzhou in Guangxi province to avoid drinking water from the river. Although local fire officials dumped hundreds of tons of neutralizer into the river to dissolve the cadmium, many fish still died.

According to the research report of relevant experts, the financial loss caused by water pollution in China accounts for 1.5% to 2.8% of GDP. Humans need to adapt to their environment, not vice versa. Man has to learn not to damage nature to the point of being irretrievable. In order to save costs, many Chinese factories discharge waste liquids directly into rivers, lakes, and streams without treatment. The laxity of law enforcement officials is another factor contributing to the further escalation of water pollution. This serious environmental pollution is an illegal act accumulated over the years. Long term use of chemical fertilizers and pesticides by farmers pollutes the land. Other common sources of pollution are surface waste dumps, landfill sites, tailings storage, etc. which aggravate the deterioration of underground water contamination.

11.8 Dianchi Lake in Kunming Seriously Polluted

The Dianchi Lake in Kunming, Yunnan province, is one of the most polluted in China. Covering an area of 330 square kilometers, Dianchi Lake is the largest fresh water lake in Yunnan Province. It is known as the pearl of the Plateau with beautiful scenery and is a national tourism resort in China. According to a report, the Yunnan provincial government has so far invested more than 60 billion yuan to control the pollution of Dianchi Lake. After years of pollution control, although the environmental deterioration of Dianchi Lake has been alleviated to a certain extent, pollution still remains. Kunming's population produces more than 100 million cubic meters of domestic sewage every year, pouring into Dianchi Lake. Dianchi Lake has become a dead lake. Widespread pollution has poisoned the fish in Dianchi Lake. The lake water is not fit for drinking.

Dianchi Lake, located in the lower reaches of Kunming city, is at the lowest area of Kunming city, so it has unfortunately become the discharge place for industrial wastewater and domestic sewage. The rainy season of the Dianchi Lake basin is concentrated here. A heavy rain will carry all the sewage and dirt on the ground into the Dianchi Lake, aggravating the pollution problem. Residence waste products also enter the Dianchi Lake. Only strict implementation of pollution prevention policies and strict laws can solve the serious pollution problem in Dianchi Lake forever. The government should immediately dig trenches around Dianchi Lake to prevent any waste water and rainwater from flowing into the lake. All wastewater and rainwater must be treated chemically and by filtration before they can discharge into Dianchi Lake. Only through effective water protection policy formulation and implementation of water pollution control can the lakes be rejuvenated.

11.9 SOUTH-TO-NORTH WATER DIVERSION PROJECT

Water scarcity has been around in China for many decades. In 1952, when Mao Zedong inspected the Yellow River, he remarked to the accompanying officials: "There is more water in the south and less water in the north, and if possible to borrow some water." A few years later, in 1959, the Chinese Academy of Sciences and the Ministry of Hydropower began to study a "South-to-North Water Diversion Plan." The effect of such a big project is far more than just transferring water and alleviating the water shortage problem. It should also revitalize the entire regional water resources through an engineering system, bringing together the Yangtze River water, the Yellow River water, local surface water, and all kinds of unconventional water to build a backbone water network of joint scheduling and optimal configuration. The project is intended to:

- Resolve water insecurity in the north
- Increase water available for irrigation, industries, and human consumption
- Provide health benefits from improved water quality
- Provide additional water to help China cope with climate change
- Reduce groundwater extraction

Construction of the project began in 2003. The project is to divert 44.8 billion cubic meters of water per year from the Yangtze River in the southern part of China to the Yellow River Basin in the arid northern part of China. The South-North Water Diversion project will require hundreds of billions of dollars of investment. The total planned area of the project affects a population of 438 million people. The total length of the planned Eastern, Central and Western diversion routes is 4,350 km. The total length of the first phase of the East and Central diversion routes is 2,899 kilometers. It is estimated that the Eastern and Central projects can transport 30 billion cubic meters of water, serving 120 million people, and bringing annual economic benefits of $14 billion to northern cities. The entire massive and complicated project may take about 40–50 years to complete. China needs to spread out the whole project in phases because of the high costs involved.

11.10 CENTRAL ROUTE OF SOUTH-TO-NORTH WATER DIVERSION PROJECT

The Central route of the South-to-North Water Diversion Project will transfer 13 billion cubic meters of water. The Central Route is divided into two parts: first, excavating a tunnel along the central and western edge of the North China Plain; and second, crossing the Yellow River through a tunnel, extending north along the west side of the Beijing-Guangzhou Railway line, with water flowing to Beijing's Summer Palace Tuancheng Lake. The South-to-North Water Diversion project focuses on solving the water shortage problem in Henan, Hebei, Beijing, and Tianjin provinces. The water-receiving cities include Nanyang, Pingdingshan, Xuchang, Zhengzhou, Jiaozuo, Xinxiang, Hebi, Anyang, Handan, Xingtai, Shijiazhuang, Baoding, Beijing, Tianjin, and 14 other large and medium cities, providing water for daily living, industrial production, and agriculture. The total area of water supply is 155,000 square kilometers. The total length of the main canal is 1277 km. The Tianjin water feeder line is 155 km long.

The Central route of the South-to-North Water Diversion Project is 1,432 kilometers long. On December 12, 2014, the Central route of the South-to-North Water Diversion Project was officially opened to the public. As of June 2020, the first phase of the Central route of the South-to-North Water Transfer project has transferred 30 billion cubic meters of water to the north in five and a half years, benefiting 60 million people along the route. According to a report, the total investment of the Central route of the South-to-North Water Diversion project was about 92 billion yuan.

11.11 EAST ROUTE OF THE SOUTH-TO-NORTH WATER DIVERSION PROJECT

The East Route will transfer 14.8 billion cubic meters of water. It is to transport water from Yangzhou Jiangdu Water Conservancy region to the northern part of China through Jiangsu, Shandong, and Hebei provinces. Construction began in 2002. The water supply is distributed in the Huaihe River, Haihe River, Yellow River basin of 25 prefecture-level cities. The characteristic of the East Route is that the construction cost is low. The cost of the first phase is only 50 billion yuan out of a total of 250 billion yuan budget. It is because the eastern route takes advantage of the Beijing-Hangzhou Grand Canal and some existing waterways, requiring less new construction. However, it has a major disadvantage in that the water level in the south is low, while the water level in the north is high, and hence many areas of the East Route need to pump water. The East Route Project has created the largest group of pumping stations in the world, with 51 pumping stations.

In 2019, northern Jiangsu suffered a rare meteorological drought in 60 years, and the major lakes such as Hongze Lake, Luoma Lake, and Nansi Lake were close to the dry water level. To this end, the South-to-North Water Diversion Project in Jiangsu Province had opened eight pumping stations. After four months of continuous operation of the South-to-North water diversion project, it had successfully completed the task of transferring water to Shandong. All pumping stations had been operating safely for 75 days, transferring 2 billion cubic meters of water per day in order to alleviate the drought in northern Jiangsu.

11.12 West Route of the South-to-North Water Diversion Project

The West Route project will transfer 17 billion cubic meters of water. It is a long-distance water transfer project from the Yalong River and Dadu River tributaries of the upper reaches of the Yangtze River in Sichuan Province, to supplement the water shortage in the upper reaches of the Yellow River basin. It is to solve the water shortages in Qinghai, Gansu, Ningxia, Inner Mongolia, Shaanxi, and Shanxi provinces. Because the West Route project is complicated, it needs to be divided into three phases. The first phase of the water transfer will be 4 billion cubic meters, and the waterline length will be 260 kilometers. The implementation of the first and second phases of the project can meet the water resources needs of the six provinces in the upper reaches of the Yellow River around 2030. The third phase of the project, which transfers 8 billion cubic meters of water, will take a longer time and is expected to be implemented after 30 years. Investment in the West Route is expected to be 304 billion Yuan. The project hasn't started yet. The investment in water transfer on the western route is far greater than that on the Central and East Routes.

11.13 Many Canals to be Built in China

China has built many canals in the past to manage water shortage problems. So there is no reason China cannot build more canals to solve current water shortage problems. The Grand Canal of China was built in 486 BC during the Spring and Autumn Period in ancient China. The Grand Canal opened up the major north-south transport, promoting trade and cultural exchanges between different regions. The Grand Canal consists of three parts: the Sui and Tang Canal, the Beijing-Hangzhou Canal and the Eastern Zhejiang Canal. The Grand Canal has a total length of 2,700 kilometers, spanning eight provinces and cities including Beijing, Tianjin, Hebei, and Shandong. It runs through the North China Great Plain and connects the Haihe River, the Yellow River, the Huaihe River, the Yangtze River, and the Qiantang River. It was the main artery of north-south communication in ancient China. After more than 2,000 years of continuous development and evolution, until today the Grand Canal still plays an important transportation and water conservancy function.

The Grand Canal operated for more than 500 years through the Tang and Song dynasties, until the end of the Southern Song Dynasty, when part of the canal was silted up. It was replaced by the Beijing-Hangzhou Grand Canal. In ancient times, there were no excavators to dig the canals and the operation was done by human labor. China's experience of water management by canals has sustained many thousands of years. No matter how difficult it is to overcome the dilemma of water supply shortage, China is capable of solving the crisis. History shows the following famous water conservancy projects in ancient China:

- Hong Gou Canal: The earliest man-made canal connecting the Yellow River and the Huai River in ancient times. Construction began in 361 BC and continued through the Qin, Han, Wei, Jin, Southern and Northern dynasties,

- Zhengguqu Canal: It was started by the State of Qin in the last years of the Warring States period. It was built in 246 BC and located on the north bank of the Jinghe River.

- Dujiangyan Canal: Located in Dujiangyan City, Sichuan Province, a large water diversion and flood control project on the Minjiang River, is the oldest and the only one left in the history of dam-free water diversion projects. It was built from about 256 BC to 251 BC. It is the world's oldest and only remaining water conservancy project characterized by dam-free water diversion. The project was mainly for irrigation, flood control, water transport and urban water supply.

- Lingqu: Canal: An ancient canal in China that connects the Yangtze River and the Pearl River. The 37 km long Lingqu was built in 214 BC. It was the main access route between Lingnan (now Guangdong and Guangxi) and the Central Plains for more than 2000 years. It is the most complete ancient water conservancy project in the world, together with Dujiangyan in Sichuan province and Zhengguo Canal in Shaanxi Province, as the three major water conservancy projects in the Qin Dynasty.

- Tongji Weir: Located in Lishui, Zhejiang Province, it was built in 505 AD and is the oldest large-scale water conservancy project in Zhejiang Province. It is a water conservancy project mainly catered to irrigation and storage.

11.14 NEW WATER RESOURCES AND WATER GRID

To achieve the unprecedented scale of the new water resources cycle and river network system (water grid) project, the Chinese engineering team using AI conducts a comprehensive study of the extent of water scarcity, severe drought and flood damage in every region of China over the past century. Using quantum computer weather forecasting models to study topography and water flow in detail in areas, population growth, agriculture, forestry, and industry needs, China decides on the following projects:

- The water recycle: At the estuary of each river, dams are built to stop water from flowing into the sea, and pumps are used to send the water to large inland reservoirs via new canals. The pumps are powered by energy from the offshore wind turbines. All reservoirs will be filled to three-quarters of their capacity so that there is enough room for flood control during the rainy season. Water from the reservoir can be used for irrigation, industries, and human consumption. With this flood control and drought relief system, China will be able to avoid drought and flood disasters and massive economic losses in the future.

- Build reservoirs: In each new smart city, there will be two massive man-made reservoirs to ensure an uninterrupted supply of clean water. The level of each reservoir is monitored by an AI computer control system in the cities, and by an automatic water level balancing mechanism, ensuring that the reservoirs always maintain a safe level. Water exceeding the capacity of a reservoir is discharged to other reservoirs, performing like a power grid system. The newly launched self-drive satellites in November 2024 can monitor the reservoir's operation.

- The water flow in various rivers is managed on a real time basis: The satellites can analyze weather patterns and sound alarms when torrential rain would fall. Water comes from a network of canals that carry the river water back and forth. All the new flood control reservoirs are closely connected to the canals. Each reservoir is set up with automated chemical analysis laboratory tests to collect water samples daily and check the pH level and impurities of the reservoir.

- Build a national canal and river network (water grid) : China will build an unprecedented canal/river network covering the whole country, connecting the country's rivers and canals. The canal is fed by circulating water from the river outlets across the country. The circulating water flows into each new smart city reservoir, which is adjusted by the computer system monitoring bureau according to the needs of the new smart city to maintain the safety level of the reservoirs. To prevent flooding caused by the annual overflow of the Yangtze, Yellow and other rivers, the government will build many huge reservoirs in the southern part of the country to catch and retain sudden downpours.

- This new national water supply network operates like a smart electric power grid whose function is to supply in real time the exigent requirement of power needed at any region around the country due to power failure. A once-off investment in the water grid project will ensure that the whole of China will always have an unlimited supply of clean water, and will always avoid floods and droughts.

- Dredging the riverbeds of major rivers: In order to avoid river beds from rising due to silt ccumulation and eventually causing river flooding, China will embark on an unprecedented program to dredge all the excavated silt from the river bed and send to the deserts and arid land, restoring the land for planting trees and grasses, and addressing the consequences of deforestation and desertification, and gradually restoring the ecological environment to its original state. A new massive water bed excavator will be designed

- Forest wildfire system: Forest wildfire is a common disaster in the world. China has begun to plant trees throughout the country to combat climate change. In order to prevent and extinguish forest wildfires quickly, automatic fire alarm systems are installed in strategic locations in the forest. Large water pipes are built in the forest to extinguish forest fires at any time. The water source for forest fire protection comes from the newly built reservoir. The fire protection system includes an automatic fire alarm system with automatically operated pumps that spray water from the pipes onto the burning trees.

In summer, when the temperature exceeds 30 °C, the pump automatically shoots water onto the trees to prevent fires. The newly launched self-driving satellites can detect the hot spots in the forest and sound an alarm when the hotspot temperature reaches a predetermined level. In 2021 and 2023, there were 616, 709 and 328 forest fires in China. On March 15, 2024, a forest fire broke out in Baizi village, Gala town, Yajiang County, Ganzi Prefecture, and Sichuan province. Experts have indicated that recent frequent wildfires are closely linked to climatic conditions in the country's southwestern region. Given the reduced precipitation and rising temperatures, an increased risk of forest fires is possible in the near future.

11.15 SATELLITES TO HELP FLOODING AND FOREST FIRE SURVEILLANCE

In Nov. 2024, China sent a pair of self-driving satellites into space. The satellites have been designed to autonomously maintain or change their flight paths without ground intervention. They rely on visible and infrared light, allowing them to penetrate clouds, fog, and darkness, as well as capturing high-resolution images regardless of weather or time of the day or night. The satellites are designed to provide a continuous stream of high-resolution radar images of the Earth's surface. This will revolutionize surveying and mapping in the space sector.

The satellites will provide real time high resolution data all-day, all-weather radar imagery. This data is critical for various natural resource management, urban safety monitoring, emergency response of disaster warning, flood and forest fire surveillance. These features enable the monitoring of natural disasters such as earthquakes and floods, tracking deforestation, and carrying out military surveillance. In the event of natural disasters, the satellites can quickly provide images of affected areas, helping authorities assess damage and coordinate relief efforts. China aims to establish a network of at least 28 satellites to provide comprehensive Earth observation data for various applications, including natural disaster prevention. It is a masterpiece of a water grid to control water resources for living sustenance and nature disaster prevention. The satellites can monitor the US warships movement in the South China Sea.

11.16 WATER POLLUTION PREVENTION MEASURES -- ISOLATION OF POLLUTED WATER

In order to permanently prevent water pollution in rivers, lakes and oceans, China would implement the following projects:

- Digging channels beside each existing river and around each lake: All external sources of surface water and farm runoff are discharged into these channels and then diverted to the waste water treatment plant. Then the treated water flows into the reservoirs. This quarantine measure absolutely keeps rivers and lakes unpolluted. The new channel excavation uses a newly invented AI automatic control excavator. The new smart city will have a collection system that directs wastewater from homes, apartments, hotels, hospitals, offices, schools, restaurants, and supermarkets, and entertainment centers to wastewater treatment plants. The treated water will be analyzed to ensure that it meets hygienic standards before being discharged into the reservoir. After the toxic substances are removed, the remaining organic material from the sewage treatment plant will be used as fertilizer for planting. The wastewater treatment and recycling system is designed to meet the international ISO 13.060 standard.

- Livestock farm wastewater discharge control and treatment system:
Livestock farms will be limited to purpose-built ecological farms. In order to prevent soil and underground river pollution caused by animal waste, every ecological livestock farm must comply with international ISO 13.060 pollution control and management measures. Waste water

is treated to meet the standard before it is discharged into the reservoir. Factory pollution control, monitoring, and treatment systems need to comply with ISO9001 standards to ensure that total containment of factory industrial waste does not contaminate clean water sources. All factory locations are limited to a specific area.

In order to prevent disasters caused by industrial waste and chemical spills, channels will be built throughout the industrial park to surround the area in order to collect and contain any chemical spills and detergent cleaning liquids. All factories dealing with potentially carcinogenic toxic chemicals must build underground ditches to prevent the toxic elements from contaminating the environment. The factory Quality Assurance Manager must submit a monthly report to the authority, the results of the water quality control compliance test. Every factory and warehouse has an underground storage reservoir for timely fire suppression.

On the evening of August 12, 2015, a massive fire and explosion occurred at a warehouse in China's Tianjin Port, killing 173 people, injuring more than 800 others, and affecting more than 100,000 households. If this warehouse had an underground storage reservoir to extinguish the fire immediately, it could have avoided the fire spreading for a long time, causing serious damages and life loss. The newly launched self-driving satellites could have spotted the hotspot and sounded the alarm early for quick relief to stop fire spreading.

CHAPTER 12 FOOD SUPPLY IN CHINA 12.1 CURRENT FOOD SUPPLY SITUATION IN CHINA

Edible plants are essential for human and animal sustenance. Grains are consumed as foods for human and domesticated animals. The photosynthesis of plants is the absorption of energy from the sunlight, and plants absorb minerals from the soil. Because of the existence of plants, human beings on Earth can survive and multiply. It has been calculated that the entire world's green plants can produce about 400 million tons of protein, carbohydrates, and fats per day. At the same time, plants also release nearly 500 million tons of oxygen into the air, providing sufficient oxygen for people and animals. Plants not only can absorb a lot of minerals through the roots, but also transport them to the stem, leaves, fruits, and seeds. The main function of the roots of plants is to absorb water and elements. Eleven elements are needed for plant growth: carbon, hydrogen, oxygen, nitrogen, sulfur, phosphorus, zinc, calcium, magnesium, iron, and potassium. The root serves as food, medicine, and industrial raw material. Both humans and animals depend on plants and soil resources for survival.

Although China is a vast country, the per capita effective farmland area is lower than the international average. Despite China's vast land area, it is difficult to support 20% of the world's population with only 7% arable land to grow enough crops by traditional means. In addition, some farmers are semi-illiterate and have no technical knowledge of agriculture, and their survival depends on good weather. In addition, agricultural production occasionally suffers from drought, water shortage, floods, pollution, and insect attacks. Unfortunately, China's arable land continues to shrink due to expanding deserts, soil pollution, and more land being used for industrial and social infrastructure.

Population growth exacerbates the crisis of food scarcity. In February 2009, northern China experienced a severe drought which affected 141 million mu of wheat in Henan, Anhui, Shandong, Hebei, Shanxi, Gansu, Shaanxi, and other agricultural provinces. The drought also deprived 1.94 million livestock of drinking water. If China's agricultural industry continues to operate in its current traditional way, it is unlikely that China will ever be able to produce enough food to meet the country's needs, not to mention the increasing consumption of the growing middle income population.

According to reports, many farmers suffer substantially from grain damage due to attack by insects, mold, and mice and grain contaminated by soil heavy metals. In China, small-scale farming is basically unprofitable. It is a labor-intensive input and low-value production industry with high prices of fertilizers and pesticides. The quality of grain is poor, and the price is low. According to a report, in 2024, the contribution of agriculture to the Chinese GDP was about 6.8%. Agriculture employs 22.8% of the country's workforce, or approximately 177 million people. From the perspective of economic contribution on a per capita basis, the rural workers productivity is very insignificant. There is a waste of manpower and the opportunity cost is very high. There is a need to upgrade their skills and improve their productivity to be on par with international standards.

China imports a large amount of grains from the US, Australia, Brazil and Argentina, not only for human and animal husbandry consumption but also for use in manufacturing industrial products. Corn is China's main food import. 25 million tons of corn was imported in 2023,

75% of which was used for livestock feed and industrial fermentation. According to the report, China's corn production is 2.9 tons per acre, compared with 4.9 tons per acre in the US, citing the fact that one-fifth of China's soil is contaminated with heavy metals. The staple food of Chinese people is rice and wheat. China's grain production is not only constrained by natural factors such as climate, arable land and water resources, but also affected by economic and social factors such as labor, technology, geopolitics, and currency fluctuations. In 2023 China imported 12 million tons of wheat from other countries

World grain trade is controlled by four multinational grain merchants. They are three American companies: Bunge, Cargill, Archer Dennis Midland and a French company, Louis Dreyfus. Because of Washington's insistence on treating China as an enemy, some American politicians have begun to talk about imposing sanctions on grain exports to China. It is also pressuring other countries to ban the sale of food to China. The current world situation is very unstable, and China's demand for imported grain will be severely hit. Because of the surge in demand for pork, China needs to import 80 percent of its soybean needs from overseas, especially from the US, because soybeans are used as feed for pigs.

When China needs to import grains from the US, the lives of the Chinese people are subject to US political stranglehold. China must take immediate action to be self-sufficient in food production. The American government provides substantial subsidies to all their farmers. Grain imported from the US is cheaper than grain produced in China. This is perhaps one reason China prefers to buy grain from the US instead of producing its own.

12.2 China's Grain Imports

According to a report, China's grain output in 2021 was about 650 million tons and then rose to a record high of 706 million tons in 2024 with a larger harvest of rice , wheat, and corn. China has 1.4 billion people, and its per capita grain consumption is only 464 kg. The US leads the world in terms of per capita grain consumption of 1,737 kg. Though China is a major grain producer, its grain imports have been rising sharply. Data show that China's food imports mainly include soybeans, corn, and wheat . China imported in 2024 13.6 million tons of corn, 11 million tons of wheat and 105 million tons of soybeans. China's demand for soybeans has soared in recent years because they are an important source of feed for livestock. In the first five months of 2024, the US was the second-largest supplier of soybeans to China after Brazil. China's soybean imports from Brazil in the first half of 2024 totaled 34.4 million tonnes. In 2025, after Trump imposed a 145% tariff on Chinese imports, China stopped importing grains from the US.

China's grain imports were also affected by the Russia-Ukraine war. The various sanctions imposed by the US on Russia have caused disruptions to the food supply chain. In 2024 China imports more wheat from Russia and Ukraine because of a lower price of $200 against $220 per ton in the US. China also imported more corn from Brazil than the US because of lower prices. Brazilian wheat farmers were also affected by US sanctions on Russian exports because Brazil could not buy Russian fertilizer. Brazil relies on Russia for 80% of its fertilizer imports, and so international wheat prices immediately rise. The intense geopolitics have caused the global food supply chain to be very unstable for some time. A failure in one part of the supply chain affects the whole chain.

Wars in the world have a direct impact on China's grain imports, leading to fluctuating prices for grains that have sparked inflation and tensions in China, affecting China's national security. Diplomatic relations between China and the US will not improve, but will continue to deteriorate. Since Trump took office, the US has severely damaged the world's supply chain.

With national survival at stake, China has no choice but to start AI-empowered agriculture immediately to become self-sufficient in food supply. Food supply must not be constricted by the US as a political weapon. At present, China has enough resources and conditions to implement an AI-empowered agriculture industry. There is no reason China should not invest $77 billion annually similar to grain import budget in AI farms to grow soybeans, corn, and wheat. This annual amount of money will be recycled within the country forever and to provide lots of jobs and save lots of foreign exchange.

12.3 FACTORS AFFECTING AGRICULTURAL PRODUCTION IN CHINA

At present, many factors affect the production of agricultural products in China: shrinking acreage, climate impacts, insufficient water resources, low productivity levels, high production costs, soil pollution, labor shortages, geopolitical influences, high prices of imported seeds and fertilizers. China is the world's largest user of chemical fertilizers and pesticides. According to reports, China's fertilizer application per unit area is 3.9 times the world average, 3.3 times that of the European Union and 4.5 times that of the US. Widespread soil pollution, especially in southern regions such as Henan Province, has prompted the government to ban the planting of 8 million acres of contaminated agricultural land until it can be restored to normalcy. It is reported that 12 million tons of grain in China are contaminated by heavy metals in the soil every year. Implementing AI farms would bypass the problem of soil pollution. Here are the problems affecting China's food production:

- Labor scarcity

In 2000, 64% of China's population was rural, but in 2021, the rural population dropped to 36%. Young people are leaving their homes in search of a better life in the big cities. Farmers work hard to grow rice, but they can only earn less than 150 yuan per mu of land. Fewer and fewer rural farmers are working the land for a living. China's increasing urbanization rate, the change of agricultural structure and the reduction of rural employment have led to a decrease in grain production. Young people prefer to work in the cities, leaving the elderly at home to farm.

By 2020, it is reported that 300–400 million rural young residents are expected to move to the cities. There would be a severe shortage of farm labor during planting and harvesting. Planting and harvesting must be done during good weather. When good weather passes, and if planting and harvesting are delayed, it would cause huge economic losses. As time goes by, the population of small scale farmers will gradually decrease, causing a drop in grain production. AI farms can solve the labor shortage problem.

• Inadequate social infrastructure in rural areas:

Poor infrastructure in the rural areas does not accord farmers reliable market forecasts of crop prices, leaving farmers guessing what to plant. Roads in many rural areas are in poor condition, forcing farmers to sell their harvests locally or to middlemen, who offer low prices. Rising costs for labor, seeds, fertilizer, and fuel have squeezed profits for many small scale farmers. Some farmers are turning to higher-priced vegetable products and abandoning low-margin grains. Hence, the supply of grain decreased, and the price went up. The cycle repeated. Unfortunately, thousands of other farmers have the same idea of switching to vegetable farming. In many rural areas, warehousing facilities, cold storage facilities and marketing platforms are inadequate, and transportation and infrastructure are poor, preventing many farmers bringing their crops from their fields to the big market in town in time to sell

• Water shortage:

There have been cases where taking water from upstream causes downstream rivers to dry up, which has led to loud complaints from people in downstream areas. Groundwater levels are falling across northern China. As aquifers are depleted and irrigation wells run dry, farmers either revert to low-yielding dryland agriculture or abandon it altogether in more arid areas. The study shows that groundwater levels are falling sharply in the North China Plain, which includes the agricultural provinces of Hebei, Shandong, Henan, Anhui, and Jiangsu. Even a relatively small and sustained decline in China's harvest could have a significant impact on food prices. This grim situation has always existed in China. The distribution of water resources is unbalanced. To avoid severe water shortages in the region of 200 million people, China has launched a 500 billion yuan project to divert water from the Yangtze River in the south to the North China Plain along three routes. Agriculture accounts for an estimated 65% to 77% of China's total water usage.

- Climate deterioration impact

Due to unpredictable rainfall, floods and droughts, climate change affects crop growth and can also be attacked by locusts and insects. Whenever harvesting or planting is affected by bad weather, Chinese crop yields are hit hard. Over the years, climate change has caused many problems for farmers because of repeated floods in the south, as well as drought, heavy snow and water shortages in northern China. When farmers have only a small plot of land and the meager income of a large family to feed, economic hardship is unbearable. According to reports, on July 21, 2021, Henan news reported that a round of heavy rainfall affected 1.24 million people in 560 towns and 89 counties of the province. 25 people died in Zhengzhou due to extreme rain, with 44,209 hectares of crops affected and direct economic losses of 65.5 billion yuan.

- Low level of productivity

It takes a small family farmer in China 58 days to produce a ton of rice, compared with less than a day and a half for large-scale farmers in the US. The traditional method of using land to grow crops on an alternating crop basis and then leaving it fallow for a few years to regenerate the soil for future crops is not very efficient in land use. With the changing world economy, the world's increased demand for food and the rapid return on economic investment, the fallow period of land has been reduced from 3 years to 2 years. This change in fallow resulted in soil degradation and decreased yields. China does not produce enough food to feed 1.4 billion people. The productivity of small scale Chinese farmers is too low, and they will never be able to meet the food demands of the population.

Most of China's rural areas remain barren, with an annual per capita income of about $400, less than a third of that in urban areas. Urban workers enjoy the state welfare social system which includes items of social security including issuance of old-age insurance, medical, unemployment, work-related injury. Farmers do not receive similar social security benefits and good schools for their children.

- Foreign crop production affects domestic prices

When farmers in the US and Australia get a good harvest, their export prices will fall significantly, affecting the price of the same crop in China. Low prices of imported grains will cause local farmers to lose money. Both Chinese and foreign commodity buyers and speculators are active in buying up crop futures. When China turns to world markets, it will inevitably turn to the US, which controls nearly half of the world's food exports. The largest US agribusiness is Cargill, with $114.6 billion in annual revenue. It has the wealth to control global grain prices. A growing number of investment banking firms make bets on commodities futures and other complex derivatives. Commodity markets are more engulfed by global finance. Their power to control the world's agricultural products is invincible.

In 2021, China imported fertilizer products from Russia, Canada, Belarus, Norway, and Israel exceeding 1 billion Yuan, with Russia accounting for the highest proportion, reaching 28%. Russia is the world's largest exporter of fertilizer. As a result of the US sanctions due to the Ukraine war, crop production in many European and other countries have suffered greatly due to the inability to buy Russian fertilizer. Fertilizer prices are also controlled by international investors. Although China's fertilizer production can meet domestic demand, the US sanctions on Russian fertilizer exports have caused fertilizer prices to rise in China, thus hitting the cost of planting for Chinese farmers.

12.4 DESERTIFICATION AND DEFORESTATION

Desertification covers more than 30 percent of China's total land area and affects the livelihoods of more than 400 million people. Desertification is occurring at an annual rate of about 3,000 square miles, five times the size of Singapore, and is estimated to cause direct economic losses of $54.1 billion. The rate of desertification is increasing with greater intensity every year. In the coming year, more arable land will be turned into desert, especially in the arid regions of northern China. The main causes of desertification are overgrazing, overdeveloped land, deforestation, urban expansion and droughts brought on by global warming, mainly due to water shortages. Once a piece of land becomes a desert, it is difficult to restore it to its original state. Once deforestation continues unabated in the countryside, during the rainy season, the topsoil is easily washed away.

Every spring, dust storms are swept up in the north by strong surface winds and spread to many cities, including Beijing. The Beijing dust originates in the deserts of Mongolia and northern China. Dust contains many harmful pollutants, such as sulfur, soot, ash, and other toxic heavy metals such as mercury, cadmium, chromium, arsenic, lead, zinc, copper, and other carcinogens. These dust storms are harmful to human health and affect wildlife. Toxic metals migrate in the food chain from fish, livestock, and plants to humans. Only with an adequate supply of water can the problem of dust storms be solved.

China's desertified land has reached 1.74 million square kilometers, accounting for 18 percent of China's total landmass of 9.6 million square kilometers. It is equivalent to the total area of nine provinces.

To reduce desertification, China employs tree-planting robots which can work 24/7, from digging to planting to watering, all in one go. The robots are managed by autonomous navigation and operation. It is reported that each planting robot can plant 1,000 trees per hour. At present, China has invested 8,000 tree-planting robots operating in the northwest desert regions, and at least 160 million trees are planted in the desert every year.

The cost of a Huawei-made tree-planting robot is about 600,000 to 800,000 Yuan. Desert planting requires saplings, seeds, nutrients, water, and other direct costs, plus equipment maintenance, involving an investment of about 2 billion Yuan. Tree planting by these robots promotes the development of the local greening industry. Northwest and Northern regions of China such as Inner Mongolia and Gansu province are facing severe desertification problems. Huawei's tree-planting robots can help address these issues.

12.5 European and American Agricultural Seeds Monopoly

In July 2016, Germany's Bayer acquired Monsanto of the US for $66 billion. Monsanto was the world's largest agricultural species subsidiary, controlling 23 percent of the global seed market. The combination of the two companies increases Bayer's patent power, which already locks up much of the world's commercially produced food supply. These patents undermine farmers' ability to use and reuse their own seeds. The loss of seed diversity also threatens biodiversity. Biodiversity helps plants and crops protect against diseases, pests and other threats such as climate change. Bayer is expected to control 23 percent of the global seed market and 24 percent of the global pesticide market. Increasing their market share increases their influence over the market and over the producers and consumers who use their products.

At present, 95% of China's broccoli seeds are imported, and 90% of carrots, onions and other seeds are imported from abroad. Many other vegetable seeds come mostly from abroad. Most vegetable seeds can only be planted once, and if they continue to grow by themselves, the yield of subsequent planted vegetables will be greatly reduced. Therefore, numerous new seeds need to be purchased every year for a new round of planting. Vegetable breeding technology is very complex, involving haploid cultivation, gene editing, breeding and other technologies. In 2020, China's agriculture-related investment is 5.5 billion Yuan, while Bayer's investment in this area is about 16 billion Yuan. China imports $500 million worth of foreign seeds each year, with vegetable seeds accounting for 60 percent of the total.

China leads the world in hybrid rice technology and does not rely on imports. Agriculture is the top priority of the national economy, and is the fundamental support of a country's national economy. If China wants to get rid of the predicament of being controlled by others on food supply, the first thing to do in agricultural development is to cultivate its own high-quality varieties. Secondly, attach importance to the cultivation of agricultural talents. Only by implementing AI agriculture can China hope to be self-sufficient in food supply. China pays about $90 billion a year to import grains, and if China can instead convert this expenditure into investment annually on AI agriculture, producing an adequate supply of grains, it will provide many jobs in this AI agriculture development. Chinese universities produce about 70,000 science and engineering PhD graduates each year, and hence there is enough talent in the agriculture advancement projects. China 's technology is ahead of many countries.

12.6 SMART AI FARMS CAN ENSURE CHINA'S FOOD SELF-SUFFICIENCY

The smart AI farm involves the following operations:

- Smart AI agriculture is a high-tech precision crop production system. The plant growth process system is developed by integrating past knowledge, latest technology, plant biology, chemistry, photosynthesis, gene editing engineering, and various fertilizers. The smart AI farm is an automated factory farm that integrates various scientific and technological processes with the entire manufacturing operation, which is a closed-loop system, managed and controlled by an AI monitoring system. Transforming the usual manual farming, agriculture into an industry managed by AI and latest technologies. It does not rely on arable land. The entire manufacturing process includes the cultivation of seeds, planting, the allocation of various nutrients required for plant growth, and the monitoring of growth progress until the plant matures and the final harvest. Each planting process has been scientifically researched, repeatedly tested, fully verified and in line with the final specification.AI farm is an integrated industry.

- Smart AI farms can produce crops three times a year, compared to two times a year on traditional manual farms. AI farm products take only days from harvest to dispatch to consumers, simplifying the entire logistics, instead of taking months when purchasing grains from overseas. In the past, the government had to go through a long process of buying grains from overseas. The process involves getting a bank to process payment, sourcing for cheapest grains on the world market, finding shipping agents for shipping the grains to various ports in China, storing the grains, and finally distributing the grains to various regions. The whole process is costly, laborious, and time-consuming.

- Smart AI farms allow planting and harvesting to be more precise, tailored to meet local consumer demands and reduce unexpected storage problems and associated costs. Smart AI farms reduce order lead time for food supply and delivery, and reduce operating costs. AI farms can produce a variety of food crops, traditional Chinese medicine herbs and animal feed.

- Smart AI farms stop using chemical fertilizer to help the environment. The farm uses organic fertilizers, also known as "farm fertilizer." Organic matter (compounds containing carbon) as fertilizer, including manure, compost, green manure, biogas fertilizer. The nutrients contained in organic fertilizers are mostly in an organic state, and through the action of microorganisms, they slowly release a variety of nutrient elements and continuously supply nutrients to the crops. The application of organic fertilizers can increase plant productivity.

12.7 AI LABORATORY RESEARCH AND EXPERIMENTS

The aim of the AI laboratory is to do research for the best cultivation process to produce the best quality grain on a large scale in the shortest possible time. The laboratory also studies the cultivation of traditional Chinese herbal plants. The lab will experiment with a variety of plant seeds, including rice, wheat, corn, soybeans, barley, and sorghum. The final grain selected for planting should contain nutrients such as iron, manganese, magnesium, phosphorus, copper, silicon, potassium, zinc, protein, vitamins, antioxidants, and fiber. Such experiments are very tedious, complex, and time-consuming. The trial will use compost containing minerals suitable for growing various grains. Compost can be made from discarded vegetables, fruits, grass clippings, leaves, hay, straw, coffee grounds, seaweed, feathers, eggshells, papers, animal waste, and the remains of plants harvested after seed removal.

For scientific experiments, the laboratory needs hundreds of thousands of clear plastic tanks to hold the different compost needed for the growth of various grain seeds. Each tank will contain different groups of compost ingredients with different compositions to grow from seedlings to fully mature plants. The key to making a good compost is the right combination of raw materials, so that the compost can be effectively broken down, resulting in a compost that is rich in both nutrients and humus. Compost is fumigated before use. Before composting the various parts of the compost material, each part needs to be fumigated separately because it contains different pests and microorganisms

According to a scientific formula, each tank of seeds in the experiment will receive a controlled amount of water, nutrients, carbon dioxide, and at various temperatures, air pressure, humidity, and light intensity at different time intervals. Every day, the growing process is videotaped, and the information is transmitted to a cloud computer for final analysis. Each grain group is examined weekly under a microscope to determine its health and growth rate. The formulations of water, nutrients, carbon dioxide and light intensity will be adjusted for maximum quality and productivity.

The grains in each group should also conform to the traditional taste (note: inorganic vegetables produced in many places no longer have the traditional taste) Some growers use a chemical to make plants grow so fast that they lose their original taste. There was a time when the taste of chicken produced in the Netherlands was fishy, because Dutch farmers used feed with fish components, and hence Dutch chickens lost the original taste of chicken. The flavor and taste of rice produced in the north and south are different. The research team can use gene editing to change the taste of grains.

To get the best seeds of each grain may take 3–4 years of experimentation. Once the best quality seeds are achieved, they will be used for planting. AI technology will be used to determine the final seeds selected for planting. Each plant produced by the experiment has a genetic code representing its genetic structure, growth rate, temperature, pressure, humidity, light and other parameters. The input to each influence parameter is different. To obtain the final optimal plant with the best nutritional value and benefits is a tedious process. With AI, automated and computerized software programs, such tedious processes become routine.

12.8 AI FARM OPERATION

AI farms do not require manual labor. The farm is run like an industrial factory, with robots performing all production activities. From planting, harvesting, packaging and final distribution are automated. Water for agriculture comes from a reservoir near the farm. All reservoir water is filtered before it reaches the farm. The farm runs on electricity generated by solar panels, and excess electricity is stored in a huge battery station that provides power at night and on cloudy days to supply the planting trough housed in an enclosure with lighting. Each farm has a specialized factory to manufacture compost mixtures suitable for the cultivation of various grains. The operation of the factory includes the planting of grain seeds in a specially prepared planting trough, which is supported by an elevated structure. All planting troughs are anchored to their precise positions by a motorized chain system.

All planting troughs are housed inside an enclosure and the seeds are grown in a controlled environment. The seeds are grown in accordance with the laboratory formula. Cameras and sensors are installed inside the enclosure to monitor plant growth. The computer control system will automatically introduce water, nutrients, and carbon dioxide gas into the enclosure regularly through tubes. The automatic environmental control system provides the appropriate air temperature, pressure, and humidity for the air entering the enclosure. During harvesting, all the troughs are moved to a factory, where the crop is automatically processed, and the final product is packed and shipped out of the factory for distribution. Plant residues and compost materials are recycled. Ai farm factories can be built on any type of land, including deserts, swamps, and arid areas.

12.9 GENETIC STUDY OF GRAINS

Gene editing is a small, controlled adjustment to an organism's existing DNA, rather than the introduction of a new foreign gene. To breed genetically modified crops, scientists extract specific genes from one organism and transfer them into another. This recombination of genes results in organisms with excellent genetic characteristics and specific desired substances. Transgenic technology can be used to breed new varieties. The taste of plants can be genetically edited or genetically modified to achieve the desired taste. In 1994, the US approved the first genetically modified food for human consumption. About 90 percent of the corn, cotton, and soybeans grown in the US today are genetically modified.

Genetically modified organisms can accelerate crop growth, increase disease resistance, increase yields, enhance environmental resilience, and increase resistance to herbicides and pesticides. According to a report in 2019, more than 18 million farmers in 29 countries planted more than 190 million hectares (469.5 million acres) of genetically modified crops. In the US, from 1985 to 2013, over 17,000 different GM crops were approved for field trials. These included corn, soybeans, potatoes, tomatoes, wheat, canola, and rice, all genetically modified to enhance traits like herbicide, insect, fungal, and drought resistance, as well as improved flavor or nutrient content.

A Chinese research team has discovered a new gene that can improve drought resistance in rice. The study found that the new gene can reduce cellular water loss by regulating water channels in cells, thereby improving the water retention capacity of rice under drought conditions. The Chinese research team compared rice plants grown under drought conditions with those grown under normal conditions to find genes that regulate drought resistance. The results showed that over expression of this gene enhanced rice resistance to drought and salt stress.

China began to collect rice varieties of the 1990s and set up a database of rice characteristic genes. In 2010, Chinese Professor Zhong Zhangmei crossbred rice with bamboo to produce new species of bamboo rice. Bamboo and rice belong to the same family of gramineous plants, and have similar genetic composition. Bamboo rice was born by combining the genes of green bark bamboo and rice. The advantage of bamboo rice is that the yield per unit area of planting is 20-30% higher than that of ordinary rice. Bamboo rice has stronger disease resistance and climate adaptability. Bamboo rice has better taste and is rich in calcium, iron, and zinc

12.10 BENEFITS OF TRANSGENIC TECHNOLOGY

B enefits of GM technology include:

- Reduce the harm of pests and diseases,
- Reduce use of pesticides, chemical fertilizers and water,
- Improve agricultural ecological environment,
- Increase crop yields.
- Reduce production costs and increase farmers' income
- Improve the added value of products.
- GM products have a certain degree of stress resistance, some biological
- Properties strengthened, improve the taste quality and nutritional value of food.

According to the International Agricultural Biotechnology Organization, the US is currently the largest producer of genetically modified soybeans. The Royal Society of Medicine, the US National Academy of Sciences, the Brazilian Academy of Sciences, the Chinese Academy of Sciences, the Indian National Academy of Sciences, the Mexican Academy of Sciences and the Third World Academy of Sciences jointly publish Genetically Modified Plants and World Agriculture, agreeing that GM technologies can be used to produce foods that are more nutritious, more stable in storage and, in principle, better able to promote health and to benefit consumers in industrialized and developing countries. So far, there has not been a single-proven food safety problem since the commercialization of genetically modified food. In July 2022, the wheat germplasm research team of Shandong

Agricultural University analyzed 25,000 strains of wheat, and did at least 300,000 experiments to find the key genes of wheat "cancer" busters to breed new disease-resistant varieties. The wheat germplasm material with Scab resistance gene created by the team has been provided to more than 60 breeding units in China.

After years of development, China has basically established a GMO breeding research and development system, becoming one of the few countries in the world with independent gene property rights and independent industrialization of GMO crops. China has established a transgenic technology system for major crops such as rice, wheat, corn, soybean and cotton, and obtained several new genes with independent intellectual property rights and a number of new genes with potential application value. The research on the cloning of rice genome and important functional genes is quite advanced in the world, laying a solid foundation for the cultivation of new crop varieties by transgenic technology. There are reportedly at least 300,000 edible plants in the world. But 65% of our foods today come from just nine main crops. So there's plenty of room for new plants.

12.11 NEW UNMANNED ROBOTIC FARMING MACHINE

In 2021, China invented a new unmanned robotic farming machine that can handle plowing, planting of seedlings and harvesting operated by a farmer using a mobile phone. The unmanned robotic farming machine is managed by using the BeiDou navigation system and adopting technologies such as the IoT, big data, 5G communication, AI, and cloud computing. There is a smart farm in Wuhan that uses an unmanned robotic farming machine to complete rice seedling transplanting in 20,000 mu of rice fields within 20 hours. The seedling planting operation has an accuracy of 2.5 cm between each seedling, thereby maximizing the use of the farming land.

A smart driverless robotic harvester can harvest and automatically offload all the harvested grains onto an accompanied truck, which then transport the grains to a granary. The whole operation can be done in a day, compared with several days in a manual farm using several workers. Presently, many farmers are faced with shortage of temporary workers during seedling planting and harvesting. It is crucial for the farmers to do transplanting of seedlings and harvesting quickly during good weather conditions. When there is a shortage of temporary farmworkers during good weather, the farmers would suffer financial loss. The government is aiding farmers around the country to adopt technology in agriculture industries. The government is also encouraging university graduates to enter AI smart farms business.

CHAPTER 13 MEAT SUPPLY IN CHINA
13.1 WEAKNESSES IN CHINA'S ANIMAL HUSBANDRY

China has many important industries being state-owned. They maintain China's national security, such as aviation, high-speed rail, highway, bridge, petroleum, shipping, telecommunications, finance, banking, electricity, steel, mining, coal, chemical, medicine and other industries. And these industries need a sound management system, sufficient resources and capital, and talent to operate to achieve the objectives. Many small scale private enterprises' livestock investment returns are not high, and hence Chinese people do not like to invest in cattle and pig farming. China relies on imports of pigs and cattle to make up the shortfall. This situation is not safe for China's food security, because the US and its allies may one day impose meat export sanctions on China.

China wants to establish state-owned livestock production enterprises to ensure China's self-sufficiency in food supply, which is a measure of national strategic security. China has enough resources, money, talent, and technology to build AI-operated cattle, pig, and chicken livestock farms. According to historical agricultural statistics, China consumes nearly 600 million pigs every year, almost half of the world's pork. In 2022, China's annual pork production was 57.94 million tons, ranking first in the world, almost half of the global output of 124.5 million tons. With a population of 1.4 billion, China needs a huge farming industry to produce enough meat to be self-sufficient. As China's economy grows rapidly, so does its appetite for meat, especially pork. China has the world's largest pork consumption per capita. In 2023 the consumption reached 42.1 kg per capita. About 54 percent of China's entire meat industry output is pork.

13.2 PIG FARMING INDUSTRY IN CHINA

The following factors affect pig farming industry in China:

- High prices of pig feeds: 80% of feeds are imported, and the high cost of imported feeds affect the profits of the pig farming industry. The imported feeds are corn and soybean. The price of soybeans is sensitive to geopolitics and weather. To the small pig enterprises, the unstable price of feeds affects the sales price and quantity of pork.

- Breeder pigs are expensive: China spends hundreds of millions of dollars each year importing breeder pigs, mainly from the USA, Denmark, and France. In 2023, the total value of China's imported breeding pigs was US$21.58 million, with an average value of more than 19,000 Yuan per head. Dependence on breeder pig imports has a great impact on the entire pig industry in China and the stability of pork prices. Although China has its own breeder pigs, including Chenghua pig, Yunnan small-ear pig, Dahua white pig, etc., the production capacity is particularly low, which cannot meet the huge domestic market demand. Quality pig breeding is a long, arduous and complicated process with large capital cost. Breeding a qualified breeder pig takes dozens of generations and more than ten years to complete and ensure no degradation. This is a big challenge for small and medium-sized pig farm operators and investors.

- High cost of pig breeding: Foreign breed pigs have the characteristics of high fertility rate, low feed to meat ratio, low breeding cycle which will give pig owners economic advantages. The quality of Chinese breeding pigs is not as good as imported breeding pigs, so the cost competitiveness of Chinese breeding pigs is not strong. The two major problems of breeding pigs and feed can not be solved by small private enterprises, but need to be solved by large state-owned enterprises. The small pig farm in China can only manage 10 pigs per worker compared with the operation of large pig breeding enterprises in Europe and the US where it is common for a worker to manage 3,000 to 4,000 pigs resulting in high productivity.

- Growth rate: The growth rate of foreign pigs is higher than that of Chinese native pig breeds. It is said that during the growth and fattening period, the average daily weight gain of Jinhua pigs and Taihu pigs of China's native pig breeds is 453 grams, and the average daily weight gain of foreign pigs is about 667 grams. Foreign pigs can reach more than 90 kg at 180 days of age, while domestic pigs require more than 180 days of age.

- Feed utilization: The average feed to meat ratio of local pigs is 3:1, but for American pigs is about 2.7:1 which means less feeding costs.

- Lean meat rate: The lean meat rate of foreign pig breeds is higher than that of Chinese native pig breeds.

- Measuring the efficiency of pig farming: One important indicator is PSY. PSY refers to the number of weaned piglets each sow can provide per year. In developed countries, PSY can reach 22–28, while in China the average figure is 17. This low figure is due to poor farming and hygienic conditions. American piglets are weaned 21 days after birth, but Chinese piglets take 5-7 extra days.

The longer weaning time means that the sow will wait longer to breed the next generation of piglets. American sows give birth faster and more often, and piglets live longer.

- Pork production and demand is cyclical: The pork prices vary in each region depending on the number of sows available in each region. The fluctuation of pork price in the market affects the number of pigs raised. When pork prices rose, many farmers produced many pigs, driving down prices. When pork prices go down, pig production goes down, causing pork prices to go up. This reactionary approach to pig farming leads to cyclical fluctuations in pork prices and unpredictable income for farmers. Therefore, pig farming is not a good investment for many small pig breeding enterprises. From 2019 to 2020, under the influence of African swine fever and other factors, China's pig output and pork production greatly declined.

- Low standard of pig farming : China's breeding density is high, and the pigs are in sub-health. They are kept in a closed environment with no sun and fresh air, little movement every day and hence pigs are of poor health. Pig farming is a very dirty industry. Pigs produce a lot of waste and gas, and disposing of waste requires a lot of capital investment. With the intensification of market competition and the increasingly strict environmental regulation, small scale pig farmers are forced to withdraw from the pig breeding market due to high costs.

- The healthcare of sows and piglets: The health of sows has a great impact on the growth and longevity of piglets, as it directly affects the newborn weight and vitality of piglets. Milk is the only food source for piglets after birth, and the growth and development of piglets will be affected by insufficient milk. Due to various factors, the health conditions of sows in small scale pig farms are less than ideal, and hence the sows do not produce sufficient milk for the piglets to grow fast.

13.3 Chinese Pig Cannot Compete with Foreign Countries

Denmark's population is only 5.4 million, but it has 28 million pigs, an average of 5.2 pigs per person. In 2024, China produced 702 million pigs representing 57 million tons of pork, with an average of 0.5 pigs per person. It is incomprehensible that China, being the world's largest consumer of pork, has so few pigs per capita compared to Denmark. Denmark is the world's largest pork exporter, exporting to more than 140 countries. The Danish Agricultural Council estimates that Denmark accounts for 23% of the world's total pork trade. The Danish pig industry is a world leader in terms of breeding, food safety, animal welfare and hygiene. Environmentally sustainable production methods are key to pig production in Denmark. Farms are inspected by the government officials for compliance with regulations. Farms are required to provide a comfortable environment for pregnant sows to live in.

13.4 China Establishes a Pig Breeding Institute

To maintain national food security, China is to establish a pig breeding research institute to perform research and development on improving pig breeding. A number of state-run AI factories will be set up to produce at least 800 million pigs annually to satisfy about 57 m tons of pork demand. With this investment, China could save at least US$6 billion worth of pork imports annually with self-sustaining pork supply. The pig breeding Institute employs experienced livestock veterinarians, high-tech, IT and professional personnel to improve breeding of pigs around the country, carrying out hybrid production lineage, developing more effective diagnostic tools and vaccines, and preventing swine flu.

The Institute studies the use of genetic modification, artificial insemination and embryo transfer methods to produce high-quality offspring for pig reproduction. Veterinarians maintain the health of pigs in farms with periodic tests of pigs for cholesterol, blood pressure, immunity, and their nutritional conditions. Domestication of the pig in China dates back at least 3000 BC. Hence, Chinese pigs have contributed to the development of world pig breeds. According to a report, pigs from China were exported to the Roman Empire to crossbreed with European pigs in the 3rd Century BC. The US imported highly prolific breeds of Taihu and Minzhu Chinese pigs in 1989 because they have large litter sizes. Sow productivity has a large impact on pork production costs.

13.5 ARTIFICIAL INTELLIGENCE PIG FARM

China is to set up many AI managed smart pig farms around the country to be self-sufficient in quality pork production. Pig farming is a dirty industry. From piglets grown to matured size, the entire breeding process generates a large amount of daily waste, which must be properly disposed of. Pigs must be properly cared for in a hygienic and clean living environment and fed nutritious food in a timely manner. Any sick pigs must be treated quickly to avoid infecting other healthy pigs. In order for the pigs to have a happy and healthy living environment, feeding, relaxation, rest, and exercise time are scheduled, and feed is monitored and sufficiently provided. The AI controlled system dispenses a predetermined amount of nutrition feed for each step of growth from piglet, growing and flattening to market weight. The stress-free environment and healthy living conditions will ensure that the farm can achieve a 100% survival rate of pig production.

The breeding facility will be equipped with temperature sensors to monitor the facility temperature. Air-conditioning is provided for summer and heating for winter period. Fresh air is circulated inside the pig domicile. The facility will be disinfected weekly. The facility complies with international health standards for livestock breeding such as ISO 9001:2000 international quality management system certification, ISO 14001:2004 International environmental management system certification and ISO 22000 international food safety management system certification.

In order to prevent unsafe meat production, every slaughtering house and processing plant must comply with ISO 9001 standards for hygienic equipment, slaughtering processes and processing plant operations and maintenance. Every slaughterhouse and processing facility must employ professional animal quarantine officers, quality controllers, and veterinarians to ensure full compliance with regulations on meat production, processing, and safety operations. This ensures that the entire production and processing history of food products can be traced, including the original production location, packaging, and storage, and that every step is transparent. Inspectors from the Ministry of Health and the Ministry of Food have access to slaughtering houses and processing plants at any time to verify that the entire facility meets national legislation and ISO 9001 conditions.

There are four main stages of pig growth and development:

- Lactation period: piglets must eat colostrum in the first 3 days of birth, and fix the teats to prevent the sow from pressing the piglets.

- Nursery period: the nursery period begins with the weaning of piglets and ends around 70 days of age. Then the piglets need to transition to independent feeding.

- Fattening period: During this period, the pigs generally eat much and grow fast, but the body is weak in disease resistance, and hence the pig pen should be regularly disinfected.

- Maintain hygienic condition: To prevent the pigs from releasing methane gas into the atmosphere, ventilation equipment allows the methane gas to be expelled and fresh air to be pumped in. Methane gas is then fed to power stations as fuel. Air filtration is installed to filter dust, insects, and other virus particles. The indoor pig feces and urine are flushed into the collection tank connected to the waste treatment plant. The slurry becomes fertilizer in agriculture farming. Wastewater after treatment is used for pig farm cleaning, fire fighting and disinfection system operation.

The latest 5G equipment and AI will operate the entire facility, managing automatic feeding, environmental control, humidity and temperature sensors, camera surveillance in every corner, pig movement, facial recognition for signs of sickness, and air quality. Information of water and feed consumption and weight are captured and stored in cloud computers. Piglets of different ages are raised in separate compartments for proper feeding. The building has automatic feed conveyors entering the feed trough at specified intervals.

China has built a 26-storey-high pig farming building, the biggest in the world, with a capacity to slaughter 1.2 million pigs a year. The building is located in Ezhou, a city in central China's Hubei Province. The cost of the building is $554 million (4 billion Yuan) with fully automatic environmental control. There are 30000 automatic precision feeding spots. Every person entering the pig's dormitory will need to go through a strict disinfection procedure. Each floor has dormitory, delivery rooms and nursery and fattening areas. The amount of feed differs for each different age group of pigs. Only 10–12 staff are required on each floor. The entire building operation is managed by a control center to monitor the entire pig living conditions. The high rise building pig farm using AI and latest breeding technology to manage the four main stages of pig growth. China will have many more such high rise building pig farms around the country. Eventually, China has sufficient supply of pork to satisfy increasing demand for pork as the middle class population is increasing.

Michael Loong

Chapter 14: Fishing Industry in China
14.1 Water Pollution Affects Fishing Industries

Fish contributes to people's health. It is rich in omega-3 fatty acids, vitamins D and B2 (riboflavin), calcium, phosphorus, iron, zinc, iodine, magnesium, and potassium minerals. Protein and vitamins. Omega-3 fatty acids help the human body maintain good health and are found in a variety of fish such as salmon, trout, sardines, herring, mackerel. The medical authority recommends eating fish at least twice a week as part of a healthy diet. People must be careful when eating fish. Some fish contain high levels of mercury (Hg), which can damage human nerves and disrupt the development of the brain and nervous system in fetuses or young children.

Mercury is a natural element found in air, water, and food. Fish with high mercury levels include sharks, rays, swordfish, barramundi, and gemfish. Cadmium (Cd) is a trace element that is highly toxic to fish. Industrial effluents containing cadmium pollute the water when discharged into oceans, rivers, and lakes. The continuous expansion of the nickel (Ni) and cadmium battery recycling industry is a major-hidden danger of cadmium pollution. Electroplating, metalworking, welding, and painting are all operations associated with cadmium exposure.

Long-term consumption of fish containing high concentrations of cadmium may lead to human health problems such as skin damage, nerve damage, skin cancer and vascular disease. Therefore, it is absolutely necessary for the government to control and ban the discharge of cadmium into the oceans and rivers. Pesticides containing metals are mixed with water in agriculture and can leak into the groundwater, which flows into rivers, oceans, and lakes. Soil pollution from landfills containing metals can seep with rainwater to contaminate groundwater. According to reports, the total output of fishing products in China in 2022 reached 66.9 million tons.

China will start dredging drains on both sides of all existing rivers. The project is to isolate clean water from polluted water. Industrial waste water must be treated before it is discharged into the reservoir. Waste water from each factory must be treated by the company before it can be discharged into the reservoir. After water treatment, the water quality must be tested and the results of chemical tests submitted and approved by the competent authorities before the treated water can be discharged into the reservoir or river.

14.2 River Isolation Policy to Avoid Water Pollution

The long-term solution to the pollution problem affecting fishing industries is to isolate the source of water pollution by not allowing polluted water from flowing into the rivers. The new water management measures are designed to completely solve the problem of water pollution. They will not only provide the public with unlimited access to clean water and eliminate diseases caused by pollutants, thereby reducing medical costs, but also help maintain the ecological environment, allowing plankton, other microorganisms, and all kinds of fish to thrive. Rivers, reservoirs, dams, lakes, and oceans produce an unlimited amount of fish species, enough to meet local fresh fish supply needs. Aquaculture production will complement whatever is lacking in daily capture from the natural environment to meet human consumption and industrial food manufacturing. In order to ensure that there is an unlimited supply of water that is not polluted, pollution control is a practical and permanent program.

14.3 STATUS OF CHINA'S FISHING INDUSTRY

The current amount of fish harvested from the sea and rivers cannot meet the needs of China's 1.4 billion people. The decrease in the supply of fish is attributed to the continuous pollution of oceans, inland rivers and lakes by acid rain, farm pesticides and fertilizers, sewerage and discharges of toxic waste from factories. A few decades ago, China's fisheries industry was rich, and fishing boats could catch a lot of fish not far from the shores. Over time, the landscape of fishing has changed dramatically. More fish is required to meet population increase. Fishermen now have to go out further from the shores and spend more time at sea to catch more fish. High seas are rough and dangerous to fishmen. Many fishermen find fishing life too difficult to be away from the shore for a long time.

Because of long-distance fishing, fishermen need to buy larger fishing boats, with refrigeration equipment and navigation equipment, and hence require heavy capital investment. The maintenance cost also increases. As global climate change affects marine ecosystems and changes in water temperatures, causing fish stocks to decrease. From May to August each year, China bans fishing vessels fishing in the Bohai Sea, Yellow Sea, and East China Sea. The ban is part of the country's efforts to promote sustainable development of fisheries and improve marine ecology. The ban affects fishermen's income. Bad weather also affects fishing.

The World Bank reports that a major reason for the decline in global fish stocks is the use of large trawlers that catch fish of all sizes, including small fry, which are then discarded, leading to the extinction of many fish species. Overfishing is another factor of fish stock decline when too many fish are caught at once, reducing the breeding populations beyond recovery. Climate change leads to the effects of extreme weather, floods, droughts, industrial and farm drainage and air pollution that further affect the quality of water in the rivers, lakes, and seas affecting the planktons and other microorganisms.

The World Bank has recommended that fisheries cut back on fishing and give the ocean a five-year break to allow fish stocks to return to healthy levels. Nearly a third of the world's fisheries are now in serious distress due to chronic and widespread overfishing. It is reported that although China has 10 million people engaged in fishing, this number has been declining every year as the decrease in catches cannot sustain operating costs, and the grim situation is worsening. At present, the fishing industry is traditional, and the descendants of fishermen have no interests in the fishing industry, and they prefer to work on land. As fish production falls, prices rise. The government will set up a national aquaculture group to ensure that people have enough fish to eat, as the aquaculture industry is related to national food security.

14.4 CHINA TO ESTABLISH A FISHING TECHNICAL TRAINING COLLEGE

Large scale fishing with large sea going boats installed with expensive sonar equipment, radar, instruments, and machineries required heavy investment of resources and capital and hence it is beyond the capability of traditional fishing companies. State-owned fishing companies have to be established with adequate resources and skill to manage a fleet of fishing boats to catch fish around the globe. The boat captain must have the skill and experience to operate the boat equipped with AI, computerized BeiDou navigation system, sonar equipment, and all the machinery installed on the boat. He must also understand weather forecasts and schools of various fish species locations. China needs to set up a fishing technical college to offer courses to train fishermen the skill and knowledge of how to operate and maintain modern fishing boats, and to understand various fish species, weather, and fishing industry.

Chinese scientists invent a new fishing net that would not catch fish smaller than 15 cm. To ensure that the fish caught is free from infections, the chemist on board the boat will take some fish samples and test for infections. China has invented an unmanned smart diving robot that can detect shoals of fish and species in the sea. The robots will transmit real time information to the boat captain to guide the fishing nets to the fishing ground. All the fishing boats operate 24/7. The daily catches at various locations are recorded and stored in the computer for onshore analyses. The data will serve as future reference and research for more productive fishing methods, techniques, updates, and plans.

China will have a map of fish stocks, densities, and species covering a vast area of nearby sea. When fish populations dwindle, the government will release tens of millions of fry each year along the northern, central and southern coasts. This will provide plenty of fish for catching at a later stage. Such methods will help supplement the natural breeding of fish. The data of daily catch stored in the cloud computer will help the government fishery department to produce an accurate assessment of where to send the fishing boat to catch fish.

Offshore fishing boat operations are managed by the Academy of Sciences' cloud computer, based on daily fish harvest data. The computer will produce the map of the various locations of shoals of fish based on the BeiDou navigation information. After the fish are caught, they are transferred to the accompanied transport boats with special functions to keep all the caught fish fresh. The boats return to shore for the shore workers to sort, weigh and record all catches. The information will be stored in a database for future fishing plans and replenishment of fry if necessary.

14.5 NEW HIGH-TECH AQUACULTURE INDUSTRY

The Chinese government decides that the only viable solution to increase fish production to meet increasing demand is to establish a state-owned high-tech aquaculture industry, using modern technology and IT to operate scientifically, and keeping water resources unpolluted as the first step. State-owned enterprises have the human, financial and technical resources to support and sustain a national project. Marine ecosystems also need a good environment for plankton and other microorganisms to thrive and for fish to reproduce naturally.

Large aquaculture farms have been established along China's eastern coast and inland rivers, lakes, and ponds. In order to achieve a high standard of fisheries, all fisheries operators and managers must pass mandatory courses on marine life, fishing technique, maintenance of fish stocks and water pollution. The new aquaculture industry applies AI, BeiDou navigation system, IT technology, automation, R&D on lowest production costs. The newly established National Fisheries Service will initiate the production of large quantities of plankton and the regular release of various fish fry to inland waters and the sea. River and seawater quality is constantly tested for the presence of toxic chemicals. River and sea temperature and trend velocity are measured and recorded.

14.6 AUTOMATION OF MODERN INLAND FISHERY INDUSTRY

China is also building an inland automated fishing industry. Unmanned fishing boats in inland rivers are equipped with AI, sonar equipment, radar, special machineries and BeiDou navigation system. These unmanned fishing boats can operate even in inclement weather conditions, especially in winter when heavy fog and strong wind are present. The unmanned fishing boats can operate 24/7 by experienced captains stationed onshore. The location of fishing is based on the AI operated underwater drones to detect schools of fish. All caught fish are kept alive. The fishing boat arrives at the fixed river port and transfers to the wharf, where workers work on species classification and weighing. The information is stored in the database as a reference for the National Fisheries Service to analyze and to design the next fishing plan and the time for release of various fish fry.

14.7 Fishing Industry in Xinjiang

In some areas of Xinjiang, there are waters with high salinity that can farm saltwater fish. There are 88 species of fish, including 46 species of indigenous fish. Xinjiang can produce salmon, large crab, Australian lobster, South American white shrimp and other high-priced aquatic products. In 2022, Xinjiang's total fishery output reached 173,000 tons. In 2023, Xinjiang Tianyun is expected to sell 6,000 tons of salmon, which can achieve an income of about 500 million yuan. According to statistics, the total output value of Xinjiang's fisheries can reach 4.2 billion yuan in 2022.

Xinjiang is too far away from the rich eastern city markets. There is no dedicated seafood air transport to reach the big cities such as Beijing, Shanghai, Nanjing, Guangdong, and Shenzhen for sales at high prices. Yet Chinese traders can fly in lobster imports from Australia and salmon from as far away as Norway to sell at home. It is an opportune time for Chinese cargo airlines to develop seafood transport between Xinjiang and the cities in the eastern province cities. Such development will accelerate the fishery industry expansion in Xinjiang.

Chapter 15 China's Public Housing Policy
15.1 Public Residence Permit

In 1951, the Chinese government launched a policy of residence permit. The aim of the policy was to prevent farmers from migrating to the cities to live when the economy was in dire straits and lack of housing. The Government needed the farmers to stay back in the villages to produce foods for the whole country with 560 million people. It aimed to avoid severe food and housing shortages that could destabilize social order and increased concerns of unemployment, riots, and crime. This policy was a temporary measure of sacrificing farmers' freedom of movement for the greater good of the country.

Mao Zedong did promise to give every farmer land to farm, build his own house, letting them live and work in peace. In the city, each government department built houses for their workers. The size of the domicile depended on the number of children each working family had. Every worker paid rent to the Government. Farmers had free land to build their houses, but their houses did not have a property title certificate and could not be mortgaged or sold to another person.

15.2 RESIDENCE PERMIT POLICY CHANGES

Since Deng Xiaoping introduced market economy policy in 1978, China embarked on an urbanization program to allow hundreds of millions of migrant workers to live and work in the coastal cities. Because private and foreign companies in the cities do not provide housing for their employees, migrant workers have to find their own accommodation. Each government urban accommodation unit is only slightly over 30 square meters, and hence giving up a few square meters of space to rent out is a difficult situation. So most migrant workers go to the city outskirts to look for larger rooms that can be rented out at cheaper rates.

Deng Xiaoping's market economy adopted capitalism with Chinese socialism character without communism. As a result of this urbanization policy, housing demand had exceeded supply and rents had risen. At the same time, urban inflation directly affected the living standards of migrant workers. After deducting all the living and transportation expenses, the monthly income of migrant workers was greatly reduced. Many older migrant workers had found it not worthwhile to work in the cities, and they began to return to their farms, working in nearby towns in their spare time. Therefore, the industries in coastal cities encountered a serious shortage of migrant workers. Many companies raised wages for the migrant workers, but problems persisted. As time passed, Chinese enterprises had to resort to automation to solve worker shortage problems. As a result, Chinese factories producing low value goods are closing down and moving to ASEAN countries where labor is cheaper.

Chinese industry is starting to move up the value chain, producing higher-value products for export. Higher-value products require higher levels of education and better training for workers. As a result, Chinese products began to compete with those from Japan, the US, and Europe. These products include computers, washing machines, air conditioners, and household appliances. The emerging industry has provided many entrepreneurs with the opportunity to become rich, and hence they have higher disposable income to buy a new house.

Many real estate developers begin to bid higher prices to get more land to build houses. With ever-increasing higher land sale costs, housing prices begin to skyrocket. Many young families cannot afford to buy their first house, even with financial help from their parents. Some luxurious housing prices in the first-tier cities of Guangzhou, Shenzhen, Shanghai, and Beijing cost more than those in New York in terms of per square meter area. With economic progress and allowing people with mobility of living in different cities, the government decides to revoke the residence permit policy.

15.3 SOARING HOUSING PRICES

Every time the government provides a piece of residential land for housing development, housing developers bid for the land at ever-increasing prices, causing housing prices to keep rising in tandem beyond the reach of many young families. As a result, local provisional governments become richer and middle-income citizens become poorer. The local governments treat the residential land as a commodity for sale to derive an income. Such a situation should only happen in capitalist countries like the USA, where the land title is transferred to the buyer. Such residential land sale should not happen in a socialist country like China, especially the residential land sale, which only has 70 years of lease from the government. The buyer only owns the building and not the land.

In fact, everyone pays high rental to the government for the use of land for a period of 70 years. For the public to enjoy a certain standard of living, the government should ensure every family has a roof over their heads, otherwise it is impossible for the parents to raise their children. The government, acting as a custodian of the land, should only charge the citizens for land development, including sewers, water, and electricity supply and road works. The government can charge a fixed annual rental for the use of the residential land. Such a policy would lower the purchasing price of housing.

Government revenue should come from taxes and not land sales. Young people who cannot afford to buy a house are now giving up marriage, leading to a continued decline of China's population. This will greatly affect the national retirement policy significantly. Such an exorbitant land price should not happen in a socialist country. The Hong Kong government experienced a serious riot in 2019 due to the exorbitant high price of housing, beyond the ability of young families to buy.

The lands available for urban housing development in big cities such as Shenzhen, Guangzhou, Shanghai, Beijing, Nanjing, Chengdu, Tianjin, Chongqing have already reached a saturation limit. The central government should have a long-term housing development plan that does not allow more industries to be established in the first-tier cities. Instead, the central government should encourage companies to set up factories and offices in second—and third-tier cities to reduce the rising cost of land in big cities.

When first-tier cities expand further, the local governments will encounter more trouble dealing with traffic jams, lack of infrastructure and social facilities to accommodate an ever-increasing population. The high price of housing is exacerbated by many speculators who buy many houses as investment, waiting for price to rise further before selling them. Such situation is similar to the US housing bubble in 2008 that led to the subprime mortgage crisis

Presently because of the high price of housing and high bank loan interest rate, coupled with a depressed economy, many houses find no buyers. There is a housing glut in many cities. Such a situation is not only holding China's economy hostage, it is also holding the financial sector hostage. Real estate is the largest capital investment in the Chinese economy. The banks' liquidity is reduced as long as the property market remains in doldrums. The middle-income family, who buys an expensive house and has to pay the mortgage with high interest rate every month, will have less disposable income for consumption. This will directly affect the national economy and the government dual circulation economy policy. The government policy on land sale indirectly causes inflation in the economy and raises the cost of living. Such a policy should be changed to reduce inflation and the cost of living for all.

This real estate policy has absorbed much of the savings of the people by the local governments and discouraged young people from getting married and having children. In 2016, the Central Government's decision to cut the banking reserve ratio led to a new round of real estate speculation in first-tier cities. People borrowed money to invest in real estate, causing housing prices to escalate further. Such speculation not only disrupts the housing market, but also creates hidden dangers for social stability in the future. Such rapid housing price escalation causes the public undue anger.

Every day, the public is actively demanding that the government take measures to curb rising house prices. Despite government measures taken to curb price rise, they are unable to restrain the deterioration of real estate speculation. Many young people prefer to stay single and live with their parents. This situation resembles Donald Trump's policy in 2025 imposing high tariffs on imported goods, causing exorbitant price increases of consumer products and raising revenue for the government. High land price is a hefty tax on the citizens of all classes.

With the effect of COVID-19 causing many factories to close down and increasing unemployment, many people defaulted on their loans until banks were forced to sell the mortgaged properties. The trade war between China and the US exacerbates the property market glut further. The housing glut has resulted in many developers suffering from bad loans valued at more than a trillion yuan. This has caused distress to all the mortgage financial banks. Hence, the capital market is devoid of these trillion yuan for circulation. Many companies related to the housing development projects also face similar fate of closing down, resulting in massive layoff of workers. This cascading financial debacle ruins the country's economy.

China's economy has taken a severe beating at a time when the world economy has shrunk due to the war in Ukraine affecting China's global trade. The majority of the population cannot bear a double whammy of high housing prices and depressed employment prospects. High housing prices have a cascading ill effect on the national economy. It is choking the circulation of spending money of the consumers, thereby strangling the domestic market economy progress.

It is an opportune time for the Central Government to revamp the current economic policy. A booming property market is like a booming stock market as the rising price cannot go up relentlessly for too long a time and soon will plunge drastically creating a mayhem in the financial market. There should be an oversight by the socialist government to rein in the uncontrolled, rampant rise of housing prices. There should be a circuit breaker to cap the rising price.

15.4 REAL ESTATE DEVELOPER EVERGRANDE GROUP FINANCIAL WOE

Signs of the current Chinese property market crash began in August 2021. Evergrande, China's second-largest property developer, had reportedly informed the Guangdong provincial government that it was running out of cash. The company, which has more than US$300 billion in debt, has been defaulting on payments. In China's previous residential and commercial real estate bubbles, average prices tripled from 2005 to their peak in 2009. In 2011, home prices in Beijing and Shenzhen surged 140 percent in five years. Residential investment as a share of China's gross domestic product rose from 2 percent in 2000 to 6 percent in 2011, the same level that the US housing market reached before it collapsed in 2008. It is unusual for Evergrande to incur such phenomenal debt without warning by its external auditors and government oversight, especially the massive debt is in US dollars. The central government normally holds a tight rein on foreign exchange transactions.

Evergrande's slow bankruptcy triggered the latest crash in China's property market in 2022, and many property developers are also facing a new debt crisis. Evergrande bankruptcy has a cascading ill effect on its thousands of suppliers' business operations. As US dollar interest rates rise, real estate companies' dollar loans are under severe pressure from the dollar exchange rate. Chinese banks have tapped the bond market to raise 30 percent more money to shore up reserves in anticipation of heavy losses on loans. US and European companies have significant exposure to Evergrande through their holdings of corporate bonds, increasing the risk of problems outside China. Problems in China's property market have a significant impact on its economy because the sector accounts for about a third of the national economy.

The collapse of China's property market has implications for the global economy as well as for China's domestic growth. Banks will have no money to lend and the country's economy will slow down. On September 28, 2023, Evergrande Group's shares were suspended from trading in Hong Kong following reports that the company's chairman, Xu Jiayin, had been placed under police surveillance.

In a sign of the depth of China's property crisis, Country Garden, the country's largest property developer, reported a 96 percent plunge in profits for the first half of 2022. The company said that in 2022, the real estate industry faced numerous challenges, including weakening market expectations, the impact of the pandemic, weak demand and falling property prices. More than 30 Chinese property companies have defaulted on international debt. The debt crisis is rippling through the banking and asset management sectors, with debt-laden developers halting construction projects. The Chinese currency has fallen to its lowest level against the dollar since the 2008 financial crisis, adding to the challenges for Chinese property developers. Chinese property developers are the country's biggest issuers of dollar-denominated bonds. A stronger dollar in 2023 increases the cost of refinancing debt in Yuan.

The collapse of China's property market has led to a rapid slowdown in the country's economy, which has been an engine of global growth for more than a decade. China's GDP growth in 2022 was down to 2.9%. By comparison, from 2000 to 2009, China's GDP grew at an average annual rate of 10.4%, and from 2010 to 2019, at an average annual rate of 7.7%. China has set a GDP growth target of 5.0% for 2023, which is already the lowest in 30 years, except during the pandemic. Over the years, China's real estate industry has entered the state of casino activity, and bubbles can occur at any time, directly dragging down the country's economic progress.

It is an opportune time for the central government to reform its current land sale policy, which is not sustainable in forging economic development. Over concentrating on land sale as a major income for the government is a precarious economic policy that can cause serious disruption to the entire national economy and social unrest. There is also an urgent need to stop a company from getting too big to fail. The debacle of Evergrande has unearthed an urgent need to keep an oversight of the integrity of all the external auditor firms. It is a wake call for the government to cap any company from getting too big to fail due to relentless expansion and acquisitions.

15.5 How Singapore HDB Policy
Solves Housing Problems

Singapore Government's Housing Development Board (HDB) was established in 1960 after Singapore attained self-government from Britain. Before independence in 1965, Singapore had one of the worst slums in the world. The HDB was set up to embark on a policy of building housing to alleviate the severe housing shortage at that time. HDB was responsible for the planning, construction and management of new residential estates in Singapore. Now more than 80 percent of Singaporeans live in 99-year leasehold homes built by HDB. Instead of selling land to HDB for housing, the government leases it to HDB for 99 years.

HDB's policies are largely based on a consensus manifesto laid out by the Government, which aims to promote the concept of homeownership, social cohesion and patriotic nostalgia for a home. The HDB housing policy is all-inclusive for progressive national social and economic development. Another goal is to hope that people with a family will not want to emigrate, and a society with a family will be more stable. In 1968, Singapore citizens were allowed to use their pension fund (Central Provident Fund) to buy their own homes. Permanent residents can also apply to buy resale HDB flats. To avoid property speculation, the HDB has restricted an entire family to owning only one HDB flat at the same time. The Singapore government housing policy is that every one will have a roof over his head.

Singapore citizens who earn more than the maximum monthly income stipulated by the HDB are not eligible to buy HDB flats. They are free to build their private houses or buy private flats. A single person who is too poor to buy HDB flats can rent HDB flats from the government and pay a low rent for a place to live. Singapore HDB building policy has helped Singapore achieve per capita GDP of $84,714 in 2023. Not only does the Singapore government not profit from the sale of HDB flats, it also provides subsidies to first-time buyers. Young Singaporeans do not need to worry about not being able to buy a flat.

Singapore, with little land mass, has the most successful housing policy in the world. HDB's prudent land use policies have contributed to Singapore's economic and social development. Despite the population growth, residents need not worry about not being able to buy a home because HDB prebuilds flats for those who are ready to buy. Although Singapore has only a small land area of 720 square kilometers (London has 1,572 square kilometers), it can accommodate 6 million people, including foreign residents. Singapore is one of the most densely populated countries in the world. Resale HDB flats are available for sale on the open market at any time. Singapore is known around the world as a livable city with gardens. Some old HDB flats have been refurbished and rebuilt with a higher quality design and architecture.

In Singapore, every registered couple is entitled to buy one HDB flat with a 99-year lease. The current price set by the government is based on development and construction costs, and is not affected by the price of private housing. The available size of a flat for purchase depends on the combined income of the married couple. The highest income couple can purchase the largest size flat. Each couple pre-books a flat a year before they get married, and they can use their Central Providence Fund to help pay off their monthly mortgage. Such a housing policy would ease the financial burden and pressure on young couples starting a new family. The government believes the policy will keep housing shortage and rising prices under control. This policy will help a young family to have extra money to spend on daily necessity, promoting economic progress, and offering young couples to have children.

The HDB provides in the estates common social infrastructures like schools, outpatient clinic, childcare center, shopping mall, food courts and bus terminal and MRT station. A car park is provided for rental to the residents. EDB also provides a flat for couples' parents to live in the same building so that they can take care of the parents. Another important safeguard of HDB flats is that the flats cannot be used as collateral for a traditional mortgage. The HDB policy is that the flat is solely meant for living and not for other purposes.

Since the inception of the HDB, Singapore never experienced a housing shortage crisis, though with occasional delay or over supply of flats due to circumstances beyond HDB control and prediction. As Singapore's economy slowly expands and citizens' income increases, the HDB begins to introduce a new housing model named Build to Order (BTO) flats that are larger, more amenities, accessibility to key transport stations and closer to the city center. BTO flats cost more to purchase. The success of HDB relies heavily on the economy that has been progressive and the continuous upgrade of Singapore industrial value chain, prudent policy of employment of foreign workers and acceptance of more permanent residents.

15.6 Lessons Learned From Housing Bubble Crisis in US and China

The world has suffered immensely from the US financial crisis in 2008. The crisis was man-made and self-inflicted. The growth of mortgage lending, unregulated financial markets, massive amounts of consumer debt, the creation of "toxic" assets, and the collapse of home prices contributed to the global financial crisis of 2008 in the US. US mortgage-backed securities, which had risks that were hard to assess, were marketed around the world, as they offered higher yields than US government bonds. Many of these securities were backed by subprime mortgages, which collapsed in value when the US housing bubble burst due to over-inflated housing prices. Many speculative homeowners began to default on their mortgage payments in large numbers starting in 2007, resulting in plunging housing prices. The massive financial losses caused the fall of the large US investment bank Lehman Brothers in 2008.

The International Monetary Fund estimated that large US and European banks had lost more than $1 trillion on toxic assets and from bad loans from January 2007 to September 2009. Well established investment banks and commercial banks in the United States and Europe suffered huge losses and even faced bankruptcy. When the housing bubble burst, many financial institutions were left holding trillions of dollars of worthless properties. Many private investors holding mortgage-backed securities around the world incurred heavy financial losses. Many economists advocate that in the future, the financial market should be regulated like the banking industry to avoid the recurrent financial market meltdown affecting the economy of the country. Speculation on over-inflated housing prices is not a safe investment.

The combined debt of Evergrande and Country Garden sum total of US$494 billion has a huge effect on the Chinese economy and financial market. The real estate crisis has dragged down everything from the job market to consumption and household wealth. In November 2023, according to a report. There were around 20 million unfinished homes across China left by Evergrande, Country Garden and other failed developers. The estimated funding gap to complete these projects was around $446 billion. Buyers of pre-paid housing from developers are facing a number of challenges, including:

- Many unfinished homes are left incomplete with no water or electricity.
- Buyers are left with mortgages on properties that don't exist
- The government does not compensate for the losses suffered by buyers

According to a report, there are 50 million empty flats around the country. The high vacancy rate also led to 4.2 trillion Yuan of bank loans deposited in vacant housing, resulting in a huge waste of financial resources and operational standstill. These assets are idling without financial contribution to the nation. There is no white knight in sight to rescue the beleaguered Evergrande and Country Garden. Nor would there be any bailout by the government or other conglomerates during this unstable economic condition around the world. Their debt woes are rippling through China's banking and asset management industries, as the country's property debt problems affect about a third of the economy. The US Fed's 2022-2024 high interest rate exacerbates China's financial crisis.

In 2025, the US high tariff imposed on China imports further exacerbates the current debacle of economic slowdown that has led to housing prices snowballing down the deep slope in all the first tier cities. The root cause of this financial debacle is self-inflicted due to the country's land sale policy. The escalating high land price is a tax on home buyers, making housing prices unaffordable. The final straw is the high interest rate that causes the borrower to default on repayment. This land sale policy defeats the country's policy of encouraging every family having a roof over their heads. This is analogous to Donald Trump 2025 high tariff on imports causing consumers to pay higher price on imported goods leading to high inflation and failure of many small and medium size companies. It is unimaginable to think that the escalating land price has no influence on the country's economy.

All the lands are under the custodian of the government to help national and social development. The government should not treat the land as a commodity to sell to any individual and company to provide an income. Who is going to pay for the land used by the farmers to plant crops for the population, or the land used for construction of highways and high speed rail? Who is going to pay for the land used for building hospitals and schools? Every citizen should not need to buy the land to live for no more than 70–90 years. He needs a roof over his head and for his family to survive, just like the farmers who need the land to plant crops to survive. Citizens should only need to pay yearly rental for temporary use of the land. During the Chiang Kai-shek administration in China, Chinese society was controlled by the landlords who owned the farm lands and charged the farmer exorbitant rent, making their lives very miserable. During poor harvesting time, when the poor farmers could not pay rental to the landlords, some farmers resorted to committing suicide to avoid payment. Increasingly unaffordable housing is one of the key social issues underpinning the unrest in Hong Kong, resulting in massive demonstrations in 2019.

China should take a leaf from the housing policy of Singapore that possesses only a tiny piece of land mass for housing. Its renown successful HDB housing projects satisfy about 80% of the population of low and middle income citizens. Even the poorest citizen has a roof over his head by renting a very low rental HDB flat. The Singapore government does not make a profit from building and selling the HDB housing to the citizens. No one is allowed to own more than one unit of HDB housing. The rich citizens can buy privately built condominiums furnished with nice gardens, swimming pools and gymnasiums.

The Singapore government occasionally introduces punitive regulations to prevent the private residential property market from getting too hot with relentless rising prices due to speculators and strong foreign buyers. Singapore housing policy helps build a country with successful economic growth, social security, political stability and a harmonious society, even if it has the notorious reputation of car owners having to pay the world's highest price for owning a car.

Chapter 16 Banking and Finance in China
16.1 What Is Money

Money is a means of exchange of goods and services. The earliest nomadic people's physical money was livestock, animal skins, seashells, pearls, and jade as physical currency to do barter trade. These commodity currencies can only be used where local people can accept them and can only be used in small transactions. People in different regions have their own acceptance systems for commodity money. Unfortunately, there was no commodity exchange in ancient times to help them determine the value of each physical commodity. Barter trade activities were performed in an open market.

With the passage of time involving numerous extensive trade activities, people in China began to use metals such as copper, silver, and gold as commodity money, but also to facilitate cross-border trade between nations. These metal monies are durable and can be used anytime and anywhere. Various Chinese dynasties used silver ingots and minted copper alloy coins as a medium of exchange. The use of silver ingots dated back to the Han Dynasty (206 BC-AD 220). At the beginning of the Song Dynasty (960–1279 CE), the government licensed certain deposit shops where people could deposit their silver coins and receive promissory notes that could be exchanged into silver coins at the licensed deposit shops. In the 1100s, Song authorities decided to take direct control of this system, issuing the world's first proper, government-produced paper money.

Banknotes were issued in the United Kingdom in 1694 by the Bank of England. Every country's central bank issues its own national currency to be used by its citizens within the country. Each and every country can use its national currency for bilateral trade as long as both parties can accept each other's national currency at an agreed exchange rate. As the value of each country's currency is based on trust bestowed by other countries, not every country's currency is acceptable as a trading currency by other countries. Currently, the US dollar, Euro, Japanese Yen and Chinese Yuan are accepted as international trading currencies. Various national currencies are traded daily at foreign exchange institutions just like trading various commodities such as crops, meat, minerals, and oil. People in different countries need to do cross-border trade to exchange goods and services. China is the world factory producing goods for export to various countries around the world. China also needs to import goods and services from other countries.

Adam Smith (1723-1790), known as the father of classic economics, explained in his Wealth of Nations book that labor is the real price of commodities, and money is the nominal price of commodities. Without the activity of people, animals, or machines, economic activity and money cannot be generated. The primitive man in the forest who had no economic activity other than hunting produced no goods for sale. The expansion of activities generates value-add in monetary terms. Productivity per unit of time determines the value of output. The use of tractors in agriculture has better productivity than the use of cattle. Velocity of money is a measure of economic activities , the faster the velocity of money circulation, the more goods and services produced in a given period of time.

The high volume of daily US dollar circulation in international trade transactions via the SWIFT generates a high volume of revenue for American financial institutions. Countries export more goods and services and generate more monetary revenue. Many countries run deficit budgets every year to stimulate economic activity and thus increase the flow of money. The US economy is so developed because its monetary cycle is much faster than the rest of the world. American consumers like to spend money, and generally they do not save money for rainy days. In China, people save a lot of money for rainy days, such as the need to pay high fees for medical treatment. Hence, money circulation within the country is throttled as consumers spend less money to promote economic activities. When money circulates slowly, the economy moves slowly. When the US interest rate rises to the roof, the cost of money increases, causing the economy around the world to slow down and reduce world trade. Money circulation around the world is curtailed.

16.2 HIGH VELOCITY CIRCULATION ACCELERATES ECONOMIC PROGRESS

Money without circulation has no economic value. Metals and oil have no value in the ground. When the Chinese government restricts the free flow of the Yuan around the world, the Yuan is inherently less valuable than the US dollar, which flows freely around the world as a global trading currency. Therefore, the Yuan is unlikely to replace the US dollar as the sole international trading currency. The Chinese government has a conservative and strict monetary policy on the Yuan. The US government treats the US dollar as a commodity that can be traded to make more money. Every country has trust in the US dollar as being a stable currency (stable value) despite the US having a national debt of US$35 trillion in 2025.

According to a report, in 2021, the US dollar accounted for about 40 percent of SWIFT's cross-border settlement, with about $5 trillion of funds flowing daily. If the US financial institutes charge a foreign exchange fee of 2-3% on each transfer from local to US, then back to local currency exchange, they would gain hundreds of billions of dollars daily. The SWIFT is a cash cow for the US treasury, raising the US GDP greatly, doing little work. When the Fed lowers the interest rate to less than 1%, many countries will quickly borrow the US dollars (fiat money) for their economic development. Hundreds of billions of US dollars will then quickly circulate around the globe, stimulating more economic activities around the world, thereby earning more money for the US.

The limited international use of the Yuan is unbelievable, as China is the world's largest trading nation. In 2024, the Yuan kept its fourth-place spot in the ranking of payment currencies, with its share of global transactions rising to 4.74 percent. In 2023, China's share of the global market was 18.75%. China's financial transactions are still restricted to America's SWIFT system of payment due to its large trading volume with the USA, EU, and other countries that still prefer to use the US dollar for trade with China. In 2022, China reportedly paid $365.5 billion for oil imports and American financial institutes would charge China $11 billion through the SWIFT system with a 3% fee for foreign exchanges. With BRICS, China can use Yuan to buy oil, thereby saving a lot of foreign exchange every year.

In the first quarter of 2024, China's trade with other BRICS members was 1.49 trillion Yuan, or about $209.7 billion, which was an 11.3% increase from the previous year. The BRICS economies account for an estimated 37.3% of global gross domestic product based on purchasing power parity, compared with 14.5% each for the United States and the European Union. Whenever BRICS members carry out bilateral trade amongst members, they will use their own currency and stop using the US dollar. Hence, the US will lose hundreds of billions of dollars in foreign exchange revenue. Using national currency for cross border trade by BRICS members, and as China is the world's largest market and manufacturer, Chinese Yuan circulation in the world will increase many folds. More international circulation of Yuan will help the Chinese economy and financial market. The US international financial services contribute a significant portion of US GDP. In May 2025 Donald Trump stopped Bank of America and JPMorgan Chase from underwriting the Hong Kong IPO of CATL, a Chinese battery manufacturer.

16.3 CHINA'S DUAL CIRCULATION POLICY

Since Trump took office in 2016, US foreign policy has changed dramatically, disrupting world supply chain activities. His foreign policy is erratic and unpredictable. During this period of uncertainty, the Chinese government has redefined its economic plan for 2021-2025, emphasizing the introduction of "dual circulation." The new economic development pattern of "dual circulation" refers to the "internal circulation" of domestic consumption and innovation, while the "external circulation" of exports and global trade continues. The "dual circulation" is a development model that highlights the importance of further tapping the potential of domestic demand.

China's economic development is becoming more and more circular within the country, which can unlock China's growth potential. The basic economic policy of the US is based on large internal circulation of domestic consumption. It is widely believed that the average American does not have more than $600 in his bank account. Americans like to spend money to enjoy life. As a result, the entertainment industry in the US is booming. The US entertainment and social media industry is reported to reach $825 billion in 2023. The share of value-add to the GDP of the United States in 2023 by manufacturing industry is 10.3% compared to the Chinese manufacturing industry contributing 26.2% share of China's GDP. China contributes 31.6% of the total global manufacturing output.

China's large capacity of consumer products manufacturing should help the internal circulation economy. Increasing internal circulation of economic activities will accelerate the circulation of money in the country. The high cost of the health care system in the country propels many people to save money for the future in case of paying several hundreds of thousands yuan exorbitant hospital bills for a surgical operation. This money saving will be kept and not used in the internal circulation. If the government can provide a low cost of Medicare to its citizens, they will have no qualms about spending money to enjoy life better. More money in circulation will spur economic growth and increase tax payment to the government.

China currently has 600 million poor farmers who are unable to contribute much to the country's economy. Once the 200 new smart cities are built to accommodate them, they will contribute economic value to the country with skills training and better education for their children. With 600 million people increasing their economic activities, China's internal circulation economy will become more active. These 200 new smart city investments will accelerate the impetus of China's economic development. Domestic consumption has always been a major focus of the Chinese economy. China has been advocating increased domestic consumption for years, and positive results are showing. This explains why China's economy can continue to grow at a high rate without being too affected by the ups and downs of the international economy due to COVID-19 and US high interest rate.

Since 2019, China's per capita GDP has exceeded $10,000. Today's economic progress was made possible by years of export trade, which generated large amounts of foreign exchange. Only when there is economic activity will money be spent in the country, and the "external circulation" and the "internal circulation" can go hand in hand. Gradually, the domestic market expands, and the "internal circulation" becomes more active than the "external circulation" of economic activity. China's domestic market is the largest in the world.

The core objective of the new "dual circulation" project is to increase economic growth and accelerate the money flow cycle. No matter whether it is domestic investment or foreign investment, it is a good investment that can promote China's economic development. At present, many foreign investors are expanding their enterprises in China. China's opening up of its financial market is not intended to ease trade frictions with the US. Instead, it is bringing in overseas capital to encourage competition and optimize China's internal financial development.

Only by strengthening the internationalization of the yuan can China ward itself off its dependence on the US dollar, so that China will not worry about possible financial sanctions imposed by the US. Whether the money comes from the US, Japan, EU or other countries, they are good money as long as the money helps the Chinese economy to grow. Similarly, good talents are good for China, whether they are local or foreigners. Competition brings out the best commercial products for the market.

16.4 Deficiencies in China's Financial Market Operations

China is the factory of the world, but the wealth created by the productive forces is not enjoyed by some Chinese people in the rural places. China's economy can be divided into two parts: one part is the export economy to meet foreign demand, and the other part is to meet the needs of the local demand including agriculture, real estate, services, and all other domestic industries. Every time the macroeconomy goes wrong, the government suppresses the growth of the domestic economy in order to protect exports, and puts foreign demand before domestic demand. The export trade of state-owned enterprises accounts for the substantial state financial support.

Many people claim that the current economy is overheating. In fact, it is only the exports that are overheating, and the domestic economy has been too tepid, partly caused by the high cost of housing. State-owned banks lend only to state-owned enterprises and have no interest in lending to small and medium-sized enterprises (SME) to promote domestic economic activity and increase the domestic circulation of money. Money should not be restricted to only promoting external trade. SMEs account for 60% of China's GDP, contributing about 50% of the country's tax revenue and providing three-quarters of all jobs in China. When they cannot get a loan from the state-owned bank for expansion and upgrade, they borrow from the shadow banking systems, which are not regulated and are risky. Prudent utilization of financial resources is essential to keep social stability.

China's large middle class is a big driver of China's economy, and their consumption spurs China's economic progress. At present, their contribution of consumption to the Chinese economy is 37%, which is far less than the 60% to 70% in European countries and the US. China's housing policy allows banks to lend money to wealthy people to buy several properties and then leave them idle for years. These idle properties do not generate economic activities. The bank loans for these idle properties should have been offered to the SMEs or the financial market. The Chinese government should restrict bank loans only to genuine housing buyers for living and not for speculation. The Chinese financial market is overly inclined to big businesses. A one-hundred-million-yuan loan to a single large SME generates fewer local economic activities than the same amount spread across thousands of small and medium-sized companies, which urgently need financial help to sustain the livelihoods of tens of thousands of workers, at least for a short period.

16.5 The Bretton Woods Agreement

In 1944, a world monetary system centered on the dollar was established as the Bretton Woods Agreement, replacing the gold standard with the US dollar as the world reserve currency. The purpose of the Bretton Woods conference was to establish a new system of rules, regulations, and procedures for the world's major economies to ensure economic stability. In theory, it is a perfect system, but it has no direct legal authority. To this end, the Bretton Woods system established the International Monetary Fund (IMF), which would monitor exchange rates and lend reserve currencies to countries with balance of payments deficits, and the World Bank (WB), which was charged with providing financial assistance for post-WWII reconstruction and economic development in less developed countries. The IMF currently has 190 member countries and has $1 trillion in reserve currencies to lend to member countries.

The IMF Special Drawing Right (SDR) is an international reserve asset created by the IMF in 1969 to supplement the official reserves of its member countries. In 2015, the Chinese Yuan met the criteria for inclusion in the SDR basket. From the end of WWII until the early 1970s, the Bretton Woods agreement meant that participating countries' exchange rates were tied to the value of the US dollar, which in turn was tied to the price of gold.

The Bretton Woods system came into full operation in 1958 with freely convertible currencies. Countries settle their balance of payments in dollars, which are exchanged for gold at a fixed exchange rate of $35 per ounce. The US is responsible for maintaining a fixed gold price and must adjust the supply of dollars to maintain confidence in the future convertibility of gold. In the 1960s, the prolonged Vietnam War generated in the US severe inflation, which led to the removal of the dollar from the gold peg.

In 1971, US President Richard Nixon ended the convertibility of the dollar to gold and hence the US violated the Bretton Woods Agreement. This is the problem of having one country's currency dominating as a world currency. With the passage of time, the world monetary system undergoes significant changes, with the US Fed becoming the de facto authority of deciding the interest rates of other currencies against the US dollar. Raising or lowering the Fed interest rate is based on US economic conditions.

The US government can print fiat money out of thin air to flood the financial market with cheap US dollars for loans around the world at interest rates decided by the Fed. When the Fed raises the interest rate very high, the US dollar will rush back to the US from other countries which will lead to sudden shortage of US dollar in other countries, causing the local currency to depreciate against the dollar, and the debtor has to pay more local currency to buy the US dollar to repay the loan. In a way, the US dollar can exert a strangle hold on other country's economies and financial systems.

The most internationally traded currencies by value based on the distribution of currencies in the global Forex market, the five countries buying/selling currencies in April 2022 were: USA 88.5%, EUR 30.5%, JPY 16.7%, GBP 12.9%, RMB 7%. The US dollar is the most favored foreign exchange currency in international transactions. The Chinese Yuan in the foreign exchange market is very small due to the Chinese government's tight restriction of the international movement of the

Yuan, even though China's economy is the world's second largest and the world's largest international trader. In 2023, China's total import trade in goods exceeded $2.56 trillion, making it the world's second-largest importer. Exports of goods totaled about $3.38 trillion. China has a trade surplus of about $820 billion.

China's economy accounts for nearly 19 percent of the global GDP. (As of 2023, global nominal GDP was approximately $105 trillion. China's nominal GDP was estimated at $19.4 trillion) Since China's trade is mainly conducted in US dollars, the Chinese payment system must rely on the SWIFT system for dollar currency transactions. Hence, China would pay tens of billions of dollars to the US financial institutions for foreign exchange fees. China will continue to use the US dollar to trade with the US, EU and other countries allied to the US, although Chinese Yuan will be used in BRICS member bilateral trade. In 2023 China – Africa trade reached US$282 billion.

The US Fed issues treasury bonds for foreign investors to buy. Many countries will buy the treasury bonds as fixed investment and reserves. For many years , China and Japan, both having large trade surpluses in US dollars, have been the largest buyers of the treasury bonds worth as much as $1 trillion each. Many foreign companies and individuals would borrow US dollars from the American financial institutes at a very low interest rate for investments in property or stocks. When the Fed suddenly raises interest rates to very high levels, the American financial institutes will demand the foreign borrowers to repay their loans. The borrowers will need to pay a higher exchange rate to buy US dollars to repay their dollar loan, as there is scarcity of US dollars in their home market.

So far, the US has raised interest rates several times in the past few years, prompting the collapse of the currencies of some countries such as Venezuela and Turkey. The US will continue to raise interest rates in the future, depending on the US domestic economic situation. Every time the dollar circulates around the world, American financial banks make easy money by the click of the keyboard. The US dollar acts like a freely available commodity for purchase and sale worldwide. The price depends on the Fed interest rate. The US dollar can help your economy but also can tank your economy. Just like water that can float a ship but also can sink it.

16.6 PETRODOLLAR IS USA ATM

The US-Saudi Petrodollar established on 8th of June 1974 was a 50-year agreement through which Saudi Arabia agreed to trade crude oil in US dollars. The agreement was based on the US agreeing to deploy troops to defend Saudi Arabia. This means that any other country wanting to buy oil from the Saudi government will need to exchange its currency into US dollars. The oil buyer has to pay a currency exchange fee to the US financial institute. Other OPEC members followed the Saudi government accepting petrol dollars for selling oil. OPEC countries thus earn a large amount of petrodollars. The OPEC countries will use their surplus petrol dollars to buy US treasury bonds. These petrodollars are then recycled in the US and international financial markets and are increasing every year. These petrodollars and international currency foreign exchange transactions are conducted through the SWIFT trade settlements, allowing the US financial institutes to charge foreign exchange fees.

The SWIFT has become the US government's ATM. It is really easy money to earn by the US financial institutes with absolute monopoly. The US government creates flat dollars as a commodity that can be exchanged for goods with other countries. The BRICS members can now transact oil settlement using their own national currency, bypassing the US dollar. The petrodollar has been a cash cow for the US and also uses petrodollar as a political tool to sanction other countries. Oil exporting countries like Iran, Russia, and Venezuela that have been sanctioned by the US are charging their oil exports in local currencies. When every country uses its own currency to settle oil trade settlements, the US financial institutes annually will lose hundreds of billions of dollars from foreign exchange fees.

16.7 US LAW PREVENTS US FINANCIAL FIRMS FROM INVESTING IN CHINA

U nder the US Holding Foreign Companies Act of 2020, Chinese companies registered in the US face a deadline of 2024 to delist. When all US-listed Chinese companies are delisted, the US would lose $2.4 trillion in stock market value. In 2021, Biden signed an executive order barring US entities from investing in dozens of Chinese companies with alleged ties to China's defense or surveillance technology industries. Major Chinese companies previously on the Pentagon list were also added to the latest list, including Aviation Industry Corporation of China (AECC), China Mobile Communications Corp, China National Offshore Oil Corporation (CNOOC), Hangzhou Hikvision Digital Technology, Huawei Technologies and SMIC. Such a sanction policy would reduce US investment in Chinese industries. American investors' money would have to find other sources to generate income. In 2024 the US economy is suffering from high inflation, and there are not many investment opportunities in Europe or Japan. Their excess money would probably go into the local stock market and real estate to speculate, or buy government bonds.

16.8 Fed Rate Hike Triggers Global Currency Depreciation

Fed interest rate hike is a game of financial deception. As long as the Fed's higher interest rates are raised to curb domestic inflation , it will lead to the economy of other countries to be curtailed . All the dollars in foreign countries would return to the US in a rush to buy treasury bonds for higher interest. The sudden rush by foreigners to buy US dollars to repay loans causes the value of their local currency to depreciate. In December 2015, Federal Reserve Chairman Janet Yellen announced that the US would raise interest rates by 25 basis points, and later the currencies of 10 countries from the Americas to Africa, Europe to Asia were depreciated. In this way, the US government has turned the world's financial markets upside down. The Indonesian rupiah, the Philippine peso, the Malaysian ringgit, the Turkish lira, the Argentine peso, the Russian ruble and the Indian rupee had all depreciated. You need more local money to exchange petrol dollars to buy oil. The moral of the story is the US dollar can help your economy but also can tank it.

In 2023, the Federal Reserve raised the interest rate up to 5% which indirectly triggered the failure of three US banks and the world economy suffered a setback. The world economy operates under the pressure of US domestic economic policy. When US inflation rises , the Fed will raise the interest rate. The US exports its inflation to other countries. The Federal Reserve has raised interest rates several times in 2022 in an attempt to curb the rise in inflation, which is due to printing too much money to support the economy affected by the pandemic that cost $5.2 trillion. The Fed's actions reportedly pushed the M2 money supply up by $6.4 trillion between March 2020 and the end of 2021. That is an unprecedented 42 percent increase in just 22 months, far more than economic growth can absorb. When interest rates in the US approaches 5%, the flight of overseas dollars back to the US would accelerate, and foreign ownership of dollar debt would be hit hard.

According to reports, since September 2022, the exchange rate of 18 national currencies in the world has fallen by more than 10% against the US dollar, and 5 currencies have fallen by more than 20%. Four of the world's top five settlement currencies all fell against the dollar, led by the Japanese yen, which fell more than 19 percent and the British pound fell to 1.1405 against the dollar, a near 37-year low since 1985. The Yuan was the strongest performer, falling about 8%. As the world's main settlement currency, the appreciation of the dollar means the depreciation of other countries' currencies, causing their purchasing power to decline. The world economy is subject to the US financial hegemonic power.

Without adequate foreign exchange reserves, many countries will lose their own purchasing power, leading to rising prices for imported energy and food. The growth rate of most consumers' wages does not match the growth rate of prices, in which case the consumption power will be weakened. This will lead to an economic crisis, and further lead to the depreciation of the exchange rate. According to the IMF, the World Bank and others, the following countries are at risk of default: Argentina, Turkey, Lebanon, Nepal, Myanmar, Pakistan, Peru, Czech Republic, Poland, Malawi, Chile, Paraguay, Egypt, Mongolia, and Laos. The cheap US dollar can help your economy to prosper, but also can tank your economy when the US interest rate shoots sky-high. Using cheap US dollars is a potential debt trap.

16.9 MONETARY POLICY

Monetary policy is the process by which a country's central bank, currency board or other competent monetary authority draws up, announces, and implements plans of action that control the amount of money in an economy and the channels for the supply of new money. Monetary policy includes the management of the money supply and interest rates and aims to achieve macroeconomic objectives such as controlling inflation, consumption, growth, and liquidity. This is done through actions such as adjusting interest rates, buying and selling government bonds, adjusting foreign exchange rates, and changing the amount of currency that banks hold as reserves. Central bank policy announcements are valid only within the credibility of the authorities responsible for drafting, announcing and implementing the necessary measures. In an ideal world, monetary authorities should be completely immune to government and political pressure, or any other decision-making body.

Indeed, the degree of government intervention in the work of the Monetary Authority can vary across the globe. It may differ from the government, judiciary, or political parties, whose role is limited to appointing key members of an authority. Obviously, every central bank will try to formulate policies to help the financial well-being of the country, but in reality, this is impossible because the central bank has no statutory power to control government policy when it is formulated.

The US Federal Reserve has lost its ability to control US inflation because the US government continues to borrow money to fund annual budget deficits and spending increases. In 2023, the Argentine government suddenly devalued its currency by 50% due to an inflation rate of 160% in 2023 caused by years of government overspending and an increase in the money supply. The country's central bank has failed to curb high inflation for years.

On September 22, 1985, the US forced Japan to sign the Plaza Accord on the grounds of resolving its trade deficit with Japan. Less than three months after the agreement was signed, the dollar quickly fell from ¥250 to around ¥200, a decline of 20%. Three years later, the dollar fell to 120 yen per dollar. The sharp appreciation of the yen and the sharp expansion of the domestic bubble led to the 30-year stagnation of the Japanese economy caused by the bursting of the real estate bubble. The US made the Japanese economy go down when Japan over relied on the US market for its exports for too long.

China had now diversified its exports to ASEAN and EU as the leading trading partners instead of the US to avoid suffering from the same fate as Japan. Hence, President Xi Jinping refused to return a call to Donald Trump in May 2025 to beg him for mercy to reduce the US high tariff of 145% on Chinese imports, which was meant to cripple the Chinese economy. However, Trump voluntarily reduces the tariff to 35% one month later to avoid massive price increases of Chinese import goods required by the consumers. The US underestimates the strength of the Chinese economy that connects to 150 countries. The US politicians forget China is the world factory. China accounts for 30% of global manufacturing output.

When the global financial crisis hit in 2008, the Chinese government moved quickly to mitigate the impact of the crisis. Starting in the third quarter of 2008, the Chinese authorities adopted a combination of proactive fiscal policy and accommodative monetary policy, launching a 4-trillion Yuan stimulus package for 2009 and 2010 in November 2008. Those efforts to support the economy during the global financial crisis prompted a surge in bank lending. In 2009, total bank lending in China reached 9,600 billion Yuan, nearly half of that year's GDP. However, much of the money lent by banks has flowed into China's stock and property markets, rather than real economic activity, which has helped the stock market partially recover from its lows in early 2009.

Even after expansionary fiscal policies, China's debt-to-GDP ratio remained below 20% at the end of 2009. In 2010, China's central government budget deficit was just 1.7 percent of GDP, compared with 8.9 percent in the US. China has undervalued its currency for years. Between 1995 and 2005, China held its currency steady at about 8.2 Yuan to the dollar, allowing China's exports to accelerate to gain momentum. China later began a steady appreciation of its currency. By 2020, the exchange rate of the Yuan to the dollar rose from 8.2 yuan to 6.5 yuan per dollar.

Every time the US Federal Reserve raises or lowers interest rates, it affects global financial markets. In 2023, the Federal Reserve interest rate rose beyond 5%, which led to the devaluation of other countries' currencies and indirectly led to the collapse of three major banks in the US. In 2024, the Yuan-dollar exchange rate was 7.237. The Chinese central bank is very prudent with its exchange rate. The Chinese government can control its monetary policy better than others because of its political system, not influenced by private companies or individuals.

16.10 CHINA INTERNATIONALIZATION OF DIGITAL YUAN

Since the Russia-Ukraine war in 2022, the US has frozen Russia's US dollar foreign exchange reserves in foreign central banks, a practice that undermines the credibility of the dollar as a foreign exchange reserve, and it will make many developing countries worry that the US may use the dollar as a weapon to expropriate their assets in the future. Such an approach would accelerate the disintegration of the dollar-based world economic order. In the first phase of this change, many countries, mainly developing countries, will use currency swaps for foreign trade to transact, replacing the US dollar as an intermediary. However, the dollar is still the main currency in their pricing mechanism. In the second stage, a new pricing mechanism will be created, replacing the dollar. The Yuan and gold are likely to become the pricing standards during this period. However, the Yuan is still not a freely convertible currency, and gold is not convenient as a transaction medium, and so it will gradually evolve into the third stage. In the third stage, a new economic order will emerge, based on a transparent and fair digital currency payment system based on international treaties. There is a worry that using another country's currency as a global trading and reserve currency would end up with the same fate as the US dollar.

A new currency would also be created in this system, the value of which would be based on the foreign exchange reserves of the countries that are members of the BRICS. Consideration could be given to a variety of physical commodities, like gold, precious metals, industrial metals, grains, and other natural resources, as the basis for a currency index. The final new economic system would be completely free of the dollar as the currency of trade. The value of the new currency remains transparent, not subject to any country's economic condition.

China can issue two kinds of yuan: domestic digital yuan and international digital yuan. The domestic digital yuan is used only within China, while the international digital Yuan is used for international trade currency and as foreign reserves. The international digital yuan is administered by an Asian central bank organization, including the exchange rate mechanism and administration, and the creation of a new international trade digital clearing system unlike the SWIFT which is outdated involving many private banks and financial institutions.

The use of the international digital yuan will reduce the cost of bilateral trade transactions. The international digital yuan can be used as a world trade currency and can be freely exchanged with other currencies at any central bank. The international digital yuan is not subject to China's domestic economic conditions. Because the international digital yuan is administered by an international organization, it cannot be used as a political tool by any country. However, for the international digital yuan to be accepted as the world trading currency and foreign reserve, it has to gain the complete trust of every country. The international digital yuan has to retain its value under any circumstances.

16.11 CENTRAL BANK DIGITAL CURRENCY (CBDC)

A central bank digital currency (CBDC) is a digital currency issued by a central bank. The value of the digital currency is similar to the value of the banknote. The two primary categories of CBDCs are retail and wholesale. Retail CBDCs are designed for households and businesses to make payments for everyday transactions, whereas wholesale CBDCs are designed for financial institutions and banks. By March 2024, the central banks of 134 countries accounting for 98% of the world's GDP were said to be in various stages of evaluating the launch of a national CBDC. These included the ECB, the UK, and the US. China's digital yuan was the first digital currency to be issued by a major economy. Six central banks have launched a CBDC: the Central Bank of The Bahamas (Sand Dollar), the Eastern Caribbean Central Bank (DCash), the Central Bank of Nigeria (e-Naira), the Bank of Jamaica (JamDex), People's Bank of China, the Reserve Bank of India and Bank of Russia.

A major problem with central bank digital currency is deciding whether the currency should be easily traceable. If it is traceable, the government has more control than it currently does. Additionally, there's a technical aspect to consider whether CBDCs should be based on tokens or accounts and how much anonymity users should have. Governments and central banks are studying CBDCs and their implications for financial inclusion, economic growth, technology innovation, and the efficiency of bank transactions. Potential advantages include:

- Technological efficiency: instead of relying on intermediaries such as banks and clearing houses, money transfers and payments could be made in real time, directly from the payer to the payee.

- Preventing illicit activity: A CBDC makes it feasible for a central bank to keep track of the exact location of every unit of the currency.

- Tax collection: It makes tax evasion much more difficult, since it would become impossible to use methods such as offshore banking and unreported employment to hide financial activity from the central bank or government. In contrast, cryptocurrencies risk undermining efforts to crack down on corporate tax avoidance.

- Combating crime: It makes it much easier to spot criminal activity, and thus put an end to it. Furthermore, in cases where criminal activity has already occurred , tracking makes it much harder to successfully launder money, and it would often be straightforward to instantly reverse a transaction and return money to the victim of the crime.

- Proof of transaction: a digital record exists to prove that money changed hands between two parties, avoiding problems inherent to cash such as short-changing, cash theft and conflicting testimonies.

- Safety of payments systems: A secure and standard interoperable digital payment instrument issued and governed by a Central Bank and used as the national digital payment instruments boosting confidence in privately controlled money systems and increases trust in the entire national payment system

- Monetary policy transmission: The issuance of central bank base money through transfers to the public could constitute a new channel for monetary policy transmission, which would allow more direct control of the money supply than indirect tools such as quantitative easing and interest rates, and possibly lead the way towards a full reserve banking system

16.12 BRICS PAY

Financial sovereignty empowers a nation to independently manage its financial resources domestically and internationally. It entails the freedom to make financial decisions and take financial actions without external influence or coercion from other countries or international entities. Hence, BRICS is to establish the BRICS Pay system. The purpose of BRICS PAY is to make international payments less costly and complicated, encouraging international cooperation between the growing number of BRICS members. However, BRICS Pay has yet to reach the final stage of implementation due to many technical issues.

BRICS Pay will feature a decentralized cross-border message system (DCMS) which operates without a central owner. It is not affiliated with the BRICS organization or any of its councils. Participants manage their own nodes with resistance to external control or interference. DCMS automatically builds transaction routes between participants, aiming to ensure reliable communication even in case that direct transmission is unavailable. Messages will be encrypted and signed with multiple encryption mechanisms available. Participants can establish currency conversion rates and transaction limits.

The BRICS Pay encompasses the collective expertise of BRICS Business Council members from various nations, including those yet to formally join BRICS+ but committed to contributing to a new global financial and payments architecture within the BRICS+ framework. To realize the BRICS Pay project, technological, financial, legal, and consulting firms are uniting to form the BRICS Pay Consortium, operating under the principles of a decentralized autonomous organization (DAO). The Consortium adheres to the regulations of each country where its members operate. It is a network-based entity without a central headquarters. BRICS Pay needs to establish a legal framework and comprehensive seamless operating system to manage successfully. Involving so many countries and languages, it is not an easy task to establish a full-fledged BRICS PAY.

16.13 CHINA DIGITAL CURRENCY ELECTRONIC PAYMENT - DCEP

Paper money is gradually being phased out by Alipay and WeChat Pay, and Chinese society is becoming cashless. The central government has also started issuing electronic payment renminbi (digital renminbi or DCEP). This DCEP is a digital form of legal tender issued by the People's Bank of China (PBC) to replace paper money. DCEP is operated by designated operating institutions and is equivalent to banknote, with controllable anonymity, protection of personal privacy, and user information security. The DCEP application only needs the use of a mobile phone without the Internet. DCEP and the RMB will coexist for a long time. In rural or remote areas where there is no network, or foreign tourists who do not hold a Chinese bank account, they can apply for a DCEP wallet for payment. The DCEP wallet has "receive money" and "transfer" functions.

Using Alipay, Wechatpay and DCEP wallets provides convenience for people's daily life without having to go to the bank to withdraw cash or deposit money. In addition, banks do not need to hire staff to count cash movements every day. The government saves a lot of resources by not printing money. Paper money was invented by the Chinese, and it is going to be eliminated by the Chinese. This is the result of human progress. Now digital numbers are used as money. DCEP also eliminates the use of bank credit cards. Illegal money transfer and corruption can be traced.

The development of DCEP in China is mainly based on the modernization of the domestic payment system, improving efficiency and reducing costs, with a focus on serving the retail payment system.

The DCEP is a two-tier system developed by the PBC, commercial banks, telecom operators and several major third-party payment institutions. DCEP has been tested in Shenzhen, Suzhou, Xiongan, Chengdu and the Winter Olympics. DCEP release mode is to adopt a two-tier operating system. The PBC does not directly issue the central bank's digital currency to the public, but instead issues the digital currency to designated operating institutions, such as commercial banks or other commercial institutions, and then these institutions will exchange the digital currency with the public.

This two-tier operating system is basically the same as banknote issuance, so it has no impact on the existing financial system. The PBC does not carry out daily payment activities. DCEP can transfer money without relying on bank accounts, support offline transactions, and have the characteristics of "payment as settlement" (tax invoice and receipt being the same). This system is conducive to enterprises such as website shopping and related parties to enjoy the convenience of payment, and at the same time, improve the efficiency of capital turnover. It also promotes money movement acceleration. DCEP allows money to be used immediately, without manual operation or authentication or verification.

The digital yuan will be the world's first state-owned currency issued by a central bank. People can use the digital yuan anywhere because it works exactly like cash. Digital yuan can be stored in a digital wallet. Using digital yuan requires a person to download an App (digital yuan App) on the phone. The government can inspect every payment on the platform. Hence, the government can eliminate counterfeiting, money laundering operations. The government also knows exactly who spends what and where, raising potential privacy concerns.

The government has more control over the digital yuan. For example, if you are a criminal, the government can prevent you from making money transactions. If the account of a corrupt official exceeds a certain limit, the Government anti-Corruption Bureau can immediately investigate the source of the deposit. Foreign visitors to China will find it much more convenient to use the digital yuan, as they can only pay in cash if they do not have a Chinese bank account or a Chinese credit card. A digital yuan could also eliminate the problem of some employers often failing to pay their employees on time using cash. The use of high technology in daily life in China is far ahead of other countries.

DCEP can be used outside of China. Although DCEP is currently only available for domestic use, in the future everyone will be able to use it to make payments abroad. International business involving China's import and export can also be traded using DCEP. For these companies, DCEP can reduce the time to settle trade transactions, reduce counterparty risk, and make it easier to settle trade between different countries using multiple different currencies. They will no longer need to use third-party currencies such as dollars or euros to complete transactions. Therefore, DCEP promotes the internationalization of digital yuan. DCEP provides an independent option for cross-border payments. DCEP is a unifying feature of a dual circular economy, which is protected from external interference.

Digital currencies can reduce issuance costs. The cost of the issuance, printing, recycling, and storage of banknotes and coins is very high. Since digital yuan is a digital currency, no transportation or security escort are involved, thus saving time and human and material resources.

16.14 INSTABILITY IN THE WORLD FINANCIAL MARKET

The Internet provides simple procedures for digital funds to be easily transferred around the world, facilitating seamless and real-time money trading operations and financial transactions. However, some transactions, which are not carefully supervised and controlled, can be completed with the click of a mouse on the network and cause banks to suffer heavy losses; Britain's Barings Bank, for example, was founded in 1762 and collapsed in 1995, because its employee, Nick Neeson, working in Singapore, cheated the company's assets by cooking the books and opened his own account to conduct transactions digitally within his own bank. There was no one to supervise his actions, which cost Barings $1.3 billion loss.

Today's stock markets are operated by superfast computers, with high-volume traders using high-speed electronic trading to arbitrage small price differences. The system has caused several major scares. On May 6, 2010, the global stock market beacon, the Dow Jones Industrial Average, plunged nearly 1,000 points in 20 minutes, wiping out nearly $1 trillion in market value, and the "flash crash" was triggered by a trader Navinder Sarao living in suburban England. He single-handedly made $70 million in the US stock market and created a drama.

The pricing of financial products is closely related to the US dollar interest rate. When the US Federal Reserve's interest rate policy is inconsistent with market expectations, financial risks would arise. The financial risks will spread rapidly due to the existence of financial derivatives, from local risks to systemic risks. The US sub-prime mortgage crisis in 2008 turned from a local risk into a global financial storm, and stock prices plummeted worldwide.

The present scale of global economic and trade struggles, currency devaluation, competition, disputes, exchange rate manipulation and US tariff war is unprecedented. The 2024 US national debt of $35 billion is rising, and one day the dollar would collapse and the US stock market would plummet, affecting the world. Toxic derivatives are still in the world's financial markets. In 2023-2024, the US high domestic inflation and associated interest rate hike seriously affect worldwide inflation, interest rate and currency depreciation leading to many countries' economy to slow down.

It is very unfortunate that every country's economy is influenced by the US domestic economic conditions. The hegemonic power of the US dollar permeates every financial system in the world. The experience of the 1997 and 2008 global financial crises and the repeated interest rate hikes in 2023 show that the global financial market is still unstable. The dollar used as a single dominant currency in world trade and as a political instrument creates instability in the global financial market . Members of BRICS are adopting a de-dollarization policy by using each national currency to perform cross-border trade, avoiding the use of the dollar.

The world adopted the US dollar as the trusted international trade and reserve currency for years, is well-entrenched in the global financial system and hence is difficult to dislodge. Many countries rely on trade with the US to support its economy and the use of the dollar as the trading currency. Hence, as long as there are countries like China and Japan trading with the US, the US dollar and SWIFT will be used.

In March 2023, three major banks in the US, Silicon Valley Bank, Signature Bank, and Silvergate Bank, suddenly collapsed within a week. There are many reasons for this event: rising interest rates of US Treasuries, high inflation, rising national debt, global energy and food crises, the war in Ukraine, the disruption of the global supply chain and the dwindling of petrodollars. All of these factors cumulatively have contributed to instability in US and global financial markets. In 2023, there was a run on Silicon Valley Bank by frantic depositors, causing it to close down when $42 billion was withdrawn from the bank in 24 hours. Silicon Valley bank had invested much of their cash in long-term federal bonds at low interest rates. To get cash to pay back to the depositors quickly , the bank resorted to forced selling its federal bonds at a huge loss.

Donald Trump starts his presidency on 20 January 2025, and he has pledged he would impose 10% to 20% across the board tariff on all imports and as high as 145% on Chinese goods. His tariff policy will increase current inflation in the US. When inflation is on the up swing, the Fed will increase interest rates to rein in money circulation. When this happens , the US dollar will appreciate and other currencies will depreciate, affecting the international economy. A single US dollar currency as a dominant global trading currency is a risk to the world economy. Every country's livelihood is hinged on the US economic policy and its national debt crisis. What happens when one day the US economy collapses because of its huge national debt it cannot pay?

China should lead the global world trade with every country in order to help them diversify their trade with the US. China should aggressively internationalize the RMB to spur global trade. China should invest more in other countries to generate more bilateral trade activities. This should reduce the hegemonic power of the US dollar.

16.15 Financial Freedom and Globalization in China

According to the IMF assessment, with the rapid expansion of financial assets and the increasing complexity of financial products in China, financial risks are rising in China. Some of China's financial institutions do not have sufficient funds to survive the worst of the financial crisis, including the housing crisis and trade war with the US. Recently, many large Chinese enterprises have taken on huge debts in overseas mergers and acquisitions, resulting in a loan repayment crisis. Too many mergers and acquisitions using US dollar high interest mortgages and high leverage now are faced with insufficient liquidity to repay debt regularly. Evergrande with more than $300 billion in huge debt shows that the government does not have a rigorous system of financial monitoring, regulation and robust public auditing.

According to news reports, China is now facing major risks, including comprehensive risks composed of financial leverage and liquidity risk, bank non-performing loan risk, shadow banking risk, local government hidden debt risk, real estate bubble risk etc. The situation is very dire. China can take a page from the rich experience of European and American financial supervision and prevention of financial risks. The Central Bank can use big data and cloud computers to program a system to monitor whether the financial, loan and liquidity risks of large enterprises are approaching a crisis level. Large companies should automatically report their short—and long-term debt payments plan.

The Central Bank should also use AI (e.g., DeepSeek) and cloud computers to program a system to monitor the ins and outs of each bank's large sum of transfer funds to prevent internal bank employees from colluding with external criminals and misappropriating bank funds for illegal activities. The Chinese government should have a circuit breaker to ensure the big public company's financial status does not lead to a potential meltdown. In addition, the government should limit any company from expanding its business beyond its core business. China's second-biggest property developer Evergrande facing liquidation has purchased a football club with investment of 100million yuan. It also invested in many other high cost projects unrelated to its property development.

China must watch out for short-term hot money (low interest) from the US entering China's stock and property markets, which could lead to a crash, as had happened in the 1997 Asian and 2008 global financial crises. At the same time, the government must be wary of large public firms that want to borrow hot money and short-term foreign capital for their long-term projects. The flight of US dollar capital during a high Fed interest rate will cause great damage to China's financial market. So the government must strictly control and limit local companies from borrowing too much American hot money.

16.16 ASIAN INFRASTRUCTURE INVESTMENT BANK (AIIB)

The Asian Infrastructure Investment Bank (AIIB) is the first multilateral financial institution initiated by China and established in 2015. It is headquartered in Beijing and has an authorized capital of $100 billion. China is the largest shareholder with 50% of the investment. The AIIB now has 105 full member countries. The purpose of the AIIB is to promote the process of building connectivity and economic integration in the Asian region, and to strengthen cooperation between China and other Asian countries and regions. Asia accounts for one third of the global economy and is the region with the most economic vitality and growth potential in the world.

Asia is home to 60% of the world's population. However, due to the limited construction funds, some countries have serious infrastructure shortages such as railways, roads, bridges, seaports, airports, and communications that affect the economic development of the region. Due to lack of human, financial and natural resources and technical skill, many Asian countries need external assistance to help build their economy. AIIB can help China build strong diplomatic and financial relations with Asian countries to deter US hegemonic influence detrimental to China's economy. Asian countries understand during the US Secretary of State visits, he only talks about China threat and military alliance whereas Chinese foreign diplomats only talk about building roads, ports and other infrastructures.

The advent of Chinese increasing global diplomacy with the establishment of the AIIB has been inspired by China's dissatisfaction with the US and Japanese control of the World Bank and Asian

Development Bank and their lack of funding for big infrastructure projects. Many Asian countries agree with China's assessment of establishing the AIIB. The US as usual expressed reservations about the transparency of the AIIB and urged other countries not to join. As the AIIB becomes the driving force of China-led economic development in Asia, it signals the decline of the US currency monopoly and its diminishing role in international politics. The US dollar is a powerful tool in international politics. BRICS members will start using their national currencies to conduct cross-border trade in 2024, avoiding the use of the US dollar and preventing the US from imposing economic and financial sanctions on other countries. The AIIB will play a more important role in future Asian economic development projects, without attaching any political conditions as practiced by WB and ADB.

AIIB provides a platform for China to contribute resources for the building of infrastructural projects like building bridges, seaports, airports, dams, hospitals, and rails for Asian countries. The founding significance of the AIIB takes into account the importance of regional cooperation in promoting sustained growth, economic and social development of Asian economies in the context of globalization, as well as enhancing the region's ability to cope with future financial crises and other external shocks. Asia has no shortage of money. Asian billionaires have a combined wealth of $1.96 trillion. This wealth could be used to finance many infrastructural projects across Asia to promote the Asian economy. The prerequisite for the success of the AIIB is that the member country political system is stable and free from violence and corruption.

16.17 Ukraine War Bolsters Sino-Russia Bilateral Financial Integration

Russian bank customers have reportedly seen a sharp increase in the use of yuan in their accounts as a result of Western sanctions on Russia's use of the dollar and euro. At Tinkoff Bank, yuan-denominated accounts have increased eightfold. Under US restrictions, many Russian companies see yuan settlement as more stable and predictable. Another Russian bank noted that many companies involved in wholesale trade were switching to yuan. Yug Aliyev, a Russian financial adviser, told Kommersant that a huge change is taking place. In the medium term, he said, the yuan "will not only replace the US dollar in settlement with China, but will also become a more reliable means of international settlement for Russian companies."

Given that Russia's economy, finance, and trade are severely sanctioned by the US and the EU, Russian imports of daily consumable products from China have significantly increased. In 2023, China's exports to Russia were $110 billion. The total bilateral trade reached $244.8 billion in 2024 which was a record high. Chinese EV dominates Russian car sales markets. Recently, China imported from Russia 108 million metric tons of crude oil, 16.5 billion cubic meters of natural gas, 94 million tons of coal and 275000 tonnes of wheat. The US proxy war initiated by Joe Biden in Ukraine was meant to cripple the Russian economy but failed to materialize. Russia GDP growth increased to 4.1% in 2024 from 3.60 % in 2023.

16.18 SWIFT COOPERATIVES

SWIFT is a global financial messaging service provider. Established to replace telex, it is a member-owned cooperative connecting more than 11,000 banking and financial institution companies in more than 200 countries and territories. Founded in 1973 and headquartered in Brussels, the organization is a cooperative company owned and controlled by shareholders representing some 3,500 companies worldwide. SWIFT is a messaging system that operates on a global network of financial institutions. It is used by banks all over the world. SWIFT does not transfer funds⊠It can only send payment order information, and payment orders must be settled through the corresponding bank accounts owned by the institutions. It allows each member institution to be assigned a unique code that identifies not only the bank name, but also the country, city, and branch.

SWIFT code refers to the identification code assigned to every bank in the world. All members are required to pay a one-time fee plus an annual fee depending on the category of membership. In addition, SWIFT makes money by charging users for the type and length of their messages. According to the US Treasury Department, SWIFT involves about $5 trillion worth of transactions a day. This payment network allows individuals and businesses to accept electronic or card payments.

In the past, all Chinese foreign trade settlements were conducted using the SWIFT system. Because the US could use SWIFT to sanction Chinese trade, China created CIPS as an alternative to avoid using SWIFT. With more than 40 years of global operation and multi-country participation, SWIFT maintains its efficiency and global credibility, and is adequately protected against cyberattacks. Any US dollar transactions made through SWIFT will go directly to US private banks and financial institutions backed by a significant amount of US dollar capital.

CIPS, which uses Chinese state banks, does not have relatively large amounts of yuan capital. According to a World Bank report, as of 2023, the US holds approximately \$145.8 trillion, about 30.8% of the global wealth total. Hence, the US dollar wields immense power in the global financial system. However, such an immense amount of wealth possessed by the US is not verifiable. There has been no physical inspection of US gold bars stored in Fort Knox since 1953. The US has printed trillions of petrodollars out of thin air in exchange with real oil from oil exporting countries.

16.19 Cross-border Interbank Payment System (CIPS)

Cross-Border Interbank Payment System (CIPS) launched by China in October 2015 offers clearing and settlement services, enabling faster and more cost-effective cross-border RMB payments. CIPS's shareholders include HSBC and Standard Chartered of the UK, Hong Kong's Bank of East Asia and Singapore's DBS Bank, Citi of the US, Australia's Australia and New Zealand Banking Group and France's BNP Paribas. By the end of 2021, CIPS reports that 1,280 financial institutions in 103 countries and territories are connected to the system. These include 30 banks in Japan, 23 banks in Russia, and 31 banks in African countries that have received yuan assistance under China's Belt and Road Initiative. While CIPS aims to operate independently in the future, it is still working closely with SWIFT to access its wider network. In 2021, the People's Bank of China effectively formed a joint venture with SWIFT to provide local network services and store information in China. In terms of scale, CIPS is tiny compared to SWIFT.

China does not have many banks stationed in every country. Chinese banks are state-owned, and their financial regulation is conservative and not fully transparent. Since 2019, when the US threatened to decouple its economy from China, the CIPS system has become more prominent. However, according to the Bank of East Asia, most experts say that China does not have the ability to completely replace SWIFT for cross-border payments. SWIFT reportedly sends 50 million messages involving $5 trillion settlements per day, while CIPS only sends 15,000 messages per day. According to a report by the China Banking Association, the cross-border use of yuan would reach 28.38 trillion yuan in 2020.

The yuan has also reportedly moved up in the ranking of major global currencies in terms of payment volume. Data show that in January 2022, the yuan accounted for 3.2% of global payments, ranking fourth in the world. The US dollar, euro and British pound ranked the top three with 39.92%, 36.56% and 6.3% respectively. Obviously the US, EU, and UK still use SWIFT for their normal cross-border trade settlement and so do China and Japan trade with the US. CIPS is still new in the game, and it will take time to grow internationally, involving the use of English, international legal systems as well as a myriad of different countries' laws. The SWIFT has been in use for 52 years.

Chapter 17 Healthcare Services in China
17.1 China's Healthcare System

The medical healthcare standard of a country depends on many factors such as the education standard of the population, living environment, hospital medical care standard and financial resources. In 1949 when PRC was established after years of devastation of civil war and foreign invasion, China's GDP per capita was 2836 yuan and the average life expectancy was 35–40 years. The country was poor with little financial resources. Many people were illiterate. With 56 ethnic groups spread over a large land mass, it was a Herculean task to provide a good healthcare service.

However, China's healthcare service standard has improved greatly over the years. After 70 years of improved social and economic development, healthcare services in urban cities have achieved a standard on par with western countries. In 2024, China's life expectancy had increased to 79 years and GDP per capita reached US$13306. However, people living in the rural areas are still lacking in adequate medical care services. Presently, there are still about 600 million rural residents spread over the entire country. Health standard improvement is helped by better education of the population. The government has been active in reducing air pollution levels, pollution to rivers and lakes, strict regulation of industrial discharge and garbage disposal and keeping public hygiene free from infectious diseases. The cleanliness of the streets in Chinese cities are kept in tiptop conditions.

In 2023, China's health spending accounted for about 7.19% of GDP, compared with the US figure of 16% of GDP. However, this does not mean Chinese healthcare standard is lower than that of the US, as their costs of medicines are higher than China and the budget for healthcare service personnel is higher than China. China's health expenditure has been increasing consistently over the past four decades. The government has increased funding for healthcare with increased healthcare facilities and better medical technology.

Many private hospitals have been set up as supplementary to public hospital services in order to meet the increasing demands for better healthcare services. China's private hospital market has experienced rapid growth over the last decade, with private hospitals now outnumbering public hospitals by a factor of two. As of 2021, there were approximately 24,000 private hospitals in China, accounting for around 66% of the total number of hospitals in the country. This number has been steadily increasing over the years. In China, the provision and financing of health services are partly the responsibility of each local government in every province. However, more than a quarter of healthcare costs in China are co-paid by patients and the medical insurance companies.

In the past, China's public hospitals were often crowded, chaotic and noisy. The consultation rooms were scattered on different floors of older hospitals, and patients needed time to find the right place to locate a doctor. During consultation, many relatives, and friends of the patient surrounded the doctor to ask many questions, thus disturbing the doctor. Many doctors and nurses were overwhelmed and exhausted by working long hours. It has been reported that some young doctors feel work pressure too great and fall ill.

As the number of admissions increases with an aging population, the situation is getting worse. In foreign countries, minor ailments such as coughs and colds can be treated quickly in outpatient or private clinics rather than in hospitals, as is in the case in China. As many older hospitals are built in the city area, daily traffic to and from the hospital is often congested, causing many chronically ill patients stuck in traffic jams with much anxiety . The medicines dispensed by the hospital are mostly produced by domestic pharmaceutical companies.

17.2 Problems Encountered by Hospital Administration

China's economy has made great progress since the 1978 market reforms, but there are always disparities between urban and rural development, particularly in basic healthcare services. This is mainly caused by disparities in government investment, medical insurance, regional economic growth and people's income level for different regions. Also, there is the factor of vast difference in education standard between the people in the urban and rural areas. Sanitary conditions in rural areas are unsatisfactory. It is difficult for the central government to grasp the health condition of every part of the vast land of China. There is no known practical administrative policy that is proven to be effective in allocating appropriate funds, medical equipment and human resources for the healthcare for every city, town, and village equitably in a country with 1.4 billion people spread over a vast land. Aging population, illiteracy of rural population and floating population are three important factors that influence the appropriate provision of healthcare facilities.

As public hospitals are responsible for part of their annual financial budget, China's public hospitals are largely funded by sales of pharmaceutical products to the patients at a mark-up over their procurement costs. The practice may have encouraged the inflation of pharmaceutical prices. By the early 2000s, the pharmaceutical sales revenue accounted for more than 40% of total revenue of public hospitals. Patients have complained that the doctors overprescribed medicines for their treatment of minor illnesses so that the hospital can earn more income, even though the patients feel some medicines are unnecessary. Some patients also complain that they need to return to the hospital for repeat consultations.

Increasing profits from excessive prescription of expensive medicines contribute financial burden on patients and also contribute low efficiency of hospital operation. With large migration of labor from the countryside to work in the cities, the hospitals become more crowded and the doctors and nursing staff suffer from increased stresses due to overwork. Many people in the cities buy health insurance in order to help defray the hospital expenses. About 80% of the drugs sold on the Chinese domestic market are generic. Traditional Chinese medicines (TCM) are a significant component of the Chinese domestic pharmaceutical market.

17.3 Management Problems of Urban Hospitals

Many patients complain about queuing for three hours, then seeing a doctor for three minutes. There is very little communication between patient and doctor, who is working under tight time constraints. Then the doctor prescribes more medicines than necessary for minor illnesses. So patients are not satisfied by such quick consultation and this creates mistrust of the doctor. This leads to tension between doctors and patients. Sometimes the tragic death of a patient led to violent and fatal attacks on doctors and nurses by family members who were unhappy with the tragic incident.

As urban living standards rise and their knowledge of diseases and medical care increases, so does their demand for better quality medical services. At the same time, urban hospitals are also under pressure from a shortage of doctors and nurses. The influx of migrant workers to cities for treatment has also exacerbated the problem. It is no longer adequate for the current health policy to remain as is. China's current commercialized medical policy needs to change to a socialistic policy of a more affordable healthcare system.

Every winter, when smog covers northern cities, many people, especially the elderly and infants, rush to hospitals for treatment of respiratory illness. Hospitals, already overcrowded, are caught off guard by the sudden arrival of thousands of extra patients a day. This is an annual recurrent event. Conflicts between doctors and patients increase. Such a high pressure working environment would not attract many college students to choose a medical profession.

It is an appropriate time to review and analyze the shortfalls of the current legacy healthcare policy of the 1950s era to come up with a new healthcare policy to suit the 2020s social healthcare requirements. The Chinese government has advanced from paper to digital communication, using cloud and AI to keep abreast with the changing world. In the 1950s the military used cannon to fight a war and now in the 2020s uses supersonic missiles. People are now using EV instead of bicycles for transportation. People are using WeChat digital payment instead of paper money. Decades ago, patients queued for hours in the hospital to see a doctor, but nothing has changed since then. There is a disconnect in the social healthcare system.

17.4 FACTORS AFFECTING MEDICAL SERVICES

Modern hospitals need to be furnished with a host of expensive medical equipment to treat patients with different types of diseases and injuries. Equipment such as the following are needed: X-ray radiography, fluoroscopy, magnetic resonance imaging, computed tomography, hematology equipment, urine analysis equipment, infusion pump, anesthesia machine, LASIK operating machine, infusion pump, extracorporeal circulation device, respirator, electric shock generator, electrocardiogram, hemodialysis machine. Also, the hospitals need to have repair specialists to maintain all the expensive equipment. The costs of equipment, maintenance, building, and hospital staff do need a huge amount of financial resources. These costs have to be borne by the government. Fortunately , the Chinese economy has grown positively from 2836 yuan GDP per capita in 1949 to US$13,306 in 2024 and China has the second-largest economy in the world. However, these expensive Chinese hospitals are frequently visited by many patients with minor illnesses that should be treated by outpatient clinics, which are not readily available in the cities and towns.

The cost of getting a medical degree at a university is very high. Medical course duration is at least 5–6 years. After graduation, graduates are required to work in hospitals as interns for 1–2 years to gain practical experience. If a graduate wants to become a specialist, he must continue his studies and receive more medical training and guidance. Occasionally, some patients become hostile and aggressive towards some doctors whom they feel are not making their best effort to cure the patients. Therefore, not many college students like to study medicine. Nurses are always in short supply. Their jobs are tiring, and sometimes they face hostility from patients and their families.

Due to inflation , manpower shortages, rising aged population and rising salaries for medical specialists and nursing staff, it is necessary to improve the efficiency of hospital operation to keep the costs from escalating. It is necessary for the government to continuously improve the healthcare system to make hospitals more effective to treat patients and working conditions of healthcare workers. The government should be mindful of the problems of population increases and aging population affecting the working conditions of the healthcare workers and the adequacy of facilities, manpower, and equipment.

Effective communication between doctors and patients has a great impact on patient care. Many patients do not have enough vocabulary about human organs and structures to accurately describe their health condition to their doctor. Doctors are very tired after seeing too many patients at the end of the day and are unable to ask more detailed questions about the patient's symptoms and family background. Some older patients do not have a good memory of their symptoms and are unable to describe them in detail to their doctors. As a result, patients need to return to the hospital multiple times to cure their illness, resulting in higher medical costs for patients and less time for doctors to see other patients.

A good prognosis of the patient's illness depends on the doctor's proper consultation with the patient, involving the patient's daily life, working environment, food intake etc. so that correct medicine can be dispensed. A patient's history taken by a doctor is of paramount importance in the healthcare profession. In a busy, crowded, chaotic and noisy hospital environment, perfect consultation is difficult to achieve.

There is a Chinese movie entitled "I Am Dying to Survive" to reveal how a patient with a serious leukemia illness cannot afford to buy the imported original drug to cure his illness. The film exposes the story of a patient's miserable life and the shortcomings of China's policy of foreign generic drug import. The film depicts the anguish of a leukemia patient. The patient needs to pay 23,500 yuan a month to the hospital to buy a box of Swiss-made 'Gleevec' life-saving drug to stay alive. Some patients have spent more than half a million yuan to stay alive in just two years. Hence, patients who do not have the money to keep buying the drug are left to die.

One patient received information that the Indian version of 'Gleevec' generic drug was effective in treating leukemia. The Indian drug cost only 200 yuan a box. China banned the import of this Indian generic drug. One patient who imported the generic drug from India was found guilty of breaking the law. There was a wide protest against the government, which later approved the import of the Indian generic drug.

India is the largest generic drugmaker in the world. Their generic drugs are widely used in many countries, including the US. Moreover, Indian generic drugs are manufactured using active pharmaceutical ingredients sourced from China. It is time to review the policy of importing cheaper and effective generic drugs from India and other countries for the sake of the public, rather than entirely depending on domestic manufacturers. Competition brings out better quality and cheaper price goods to reduce cost of living for the public. When there is no competition, the domestic drug manufacturers monopolize the market and there is no incentive for them to reduce their cost of production.

17.5 Frequent Incidents of Doctors Being Attacked

In December 2019, Dr. Yang Wen, deputy chief physician of the emergency department of Beijing Civil Aviation General Hospital, was stabbed in the neck by a patient's son with a knife and died. The suspect was arrested. The report stated that his 95-year-old mother, after suffering a stroke, was admitted to the hospital. Despite the hospital trying to treat the patient correctly, the patient' family was angry at Dr. Yang Wen for not saving the patient's life and decided to take revenge against Dr. Yang Wen. According to reports, there have been more than 10,000 incidents of such attacks on doctors in hospitals across the country. At least 50 medical workers are reported to have lost their lives as a result of violence. A nurse in Nanjing was attacked and became paralyzed. In December 2018, a surgeon at Zhongnan Hospital of Wuhan University was stabbed by an unknown person with a knife.

According to the Chinese Medical Doctor Association, 66 percent of doctors have personally experienced doctor-patient conflicts, 51 percent have experienced verbal violence, and more than 30 percent of doctors have experienced physical violence from patients. According to to a psychologist in Chongqing, the relationship between doctors and patients has long been fraught with tension in Chinese society.

According to public opinion analysis, the patient violence incident was not only due to poor communication, medical errors, but also the patients' unrealistic expectations unfulfilled. The psychological imbalance had been due to the lack of understanding of the disease and treatment by the patients and their families. There were other causal factors related to the medical system, hospital administration, high drug prices and unequal distribution of human and financial resources.

As the population ages and hospitals are understaffed, the doctor-patient relationship will deteriorate. In the final analysis, the root problem of patient violence is due to the unsatisfactory national healthcare policy. The current hospital administration and healthcare service need overhauling. Perhaps China could take a leaf from the medical healthcare service standard in Malaysia where its high healthcare standard and affordability receives high accolades from international medical fraternity and Malaysia is not particularly a rich country. Malaysian doctor and patient relations remain cordial and non-violent.

17.6 Unnecessary High Costs of Treatments of Patients

The budget for the hospital operation comes from the taxpayers. The expensive resources spent by the hospital must have a worthy return for the taxpayers. There are no worthy return for the taxpayers treating the following patients:

- Patients lose their liver function due to alcoholism
- Patients get lung cancer due to excessive smoking
- Patients involved in drunk driving accidents leading to severe body injuries

The high costs of hospital treatment for these patients are now borne by the government and insurance companies, whose premiums are supported by all taxpayer citizens. Who should pay for their hospital expenses for these patients when their problems are self-inflicted? If there were no alcoholism, smoking and drunk driving, the hospital would save a tremendous amount of expenses. There is no worthy return of medical care investment to treat these patients. Costs in treating these patients involve the following:

- Alcohol is considered a major contributor to premature mortality in China. The British Lancet Journal survey found that China had the highest number of alcohol-related deaths in the world, with about 700,000 deaths a year directly related to alcohol, and 60 million people suffering from alcoholic liver disease like cirrhosis. Prolonged and excessive alcohol consumption can cause liver inflammation, fibrosis, and ultimately alcoholic liver cirrhosis, posing a substantial risk to liver health. The cost of treating liver cirrhosis can vary (including medications, hospital stays, and potential need for a liver transplant). Generally, the cost can be quite high, with estimates ranging from thousands to tens of thousands of US dollars per patient, with the most expensive cost being a liver transplant

- As of 2022, there are around 300 million Chinese smokers and 2.4 trillion cigarettes are sold every year, representing 46% of the world total. According to available data in 2023, an estimated over 733,000 people in China died from lung cancer, making it the leading cause of cancer deaths in the country. Lung cancer treatment depends on the type and stage of the cancer, as well as the patient's overall health. Treatments include surgery, chemotherapy, radiation therapy, targeted therapy, immunotherapy, and local treatments. A monthly dose of chemotherapy drugs can cost from US$10,000 to $12,000. The total cost of chemotherapy drugs for one year can add up to $50,000. The average cost of lung cancer surgery is around US$15,000-30,000. In 2024, the average cost of radiation treatment for lung cancer was around US$30,000-70,000 per year depending on factors such as treatment type and duration.

- Car accidents due to drunk driving can lead to a wide range of body injuries, including neck injuries like whiplash, head injuries like concussions, back pain, broken bones, internal organ damage, facial injuries, and soft tissue damage. The costs of treatment for car accidents victims are exorbitant and can cause serious loss of income to the victim's family. In 2023, the number of people killed in traffic accidents caused by drunk driving in China is as high as 100,000 each year, and it continues to rise. 30% of road traffic accidents are caused by drunk driving. 59% of driver deaths were related to drunk driving.

In order to control the escalating costs of hospital operation and to save valuable resources, China introduces a slew of policies to reduce the number of patients who are alcoholics, reckless drunk drivers and smokers. The patients will be fully responsible for payment of all their hospitalization charges. The government issues strong warnings to people not to smoke, alcohol indulgence and drunk driving. They are fully responsible for their failure to maintain a healthy life and social order.

People should understand that maintaining good health is their personal responsibility and an effective way to avoid being hit by chronic diseases. According to a survey, more than 60 million people are overweight, more than 200 million people have high blood pressure, more than 50 million people have diabetes, and 160 million people have high blood fat. It is reported that in 2020, more than 15,000 people will die every day in China due to "affluenza."

17.7 Maintain Good Health to Reduce Healthcare Costs

Prevention of disease is better than curing it. The local government decides to set up family clinics in each community to contribute to residents' healthcare services. All citizens must abide by the state's health maintenance policy. Every adult must annually go to the community family clinic for physical examination. Family clinic doctors will check everyone's blood pressure, blood sugar level, body fat, blood oxygen, urine, and heart rate and other health parameters. The government will send the physical examination results to each person, who will be notified to consult a doctor for further examination if there is any sign of health problems. The individual medical report is stored in the national medical data center for reference by a doctor in the future.

The importance of good health requires young students to be taught in school about common preventable chronic "diseases of affluence" such as diabetes, high blood pressure, cholesterol, and cancer. The school curriculum teaches students the importance of focusing on various dietary nutrients for good health needs and maintaining a good mental attitude. All government officials are forbidden to drink alcohol and smoke in official functions and work places. Tobacco or alcohol advertising is prohibited. Movies are forbidden to show scenes of smoking or drinking. The government encourages young people to develop a healthy work-life balance, balanced food intake and regular exercise. Sports are part of the school curriculum.

Understanding the causes of disease and maintaining a healthy lifestyle instilled in primary school students are core values of the new healthcare system. Let everyone have a value attached to the health service, not to waste resources. Values start from a young age to be effective, so that children develop a good habit. Maintaining physical and mental health is the most effective way to prevent the onset of preventable diseases. Early education is part of the prevention program, introducing primary and secondary school students to visit hospitals and drug rehabilitation centers to learn about chronic and preventable diseases.

Every student from primary school must participate in a sport. Health courses teach the nutritional value of different foods and the importance of a balanced diet. Teachers also take students to the supermarket to introduce various meats, vegetables, cereals and spices, and their nutritional value to human health. High school students spend part of their long summer holidays volunteering to work as nursing assistants in hospitals. They will learn about hospital care management, procedures, and equipment. They also visited the post-illness care of elderly patients in rehabilitation centers.

17.8 Anti-smoking Activities in China

China is the world's largest producer and consumer of tobacco. In 2022, China produced 2.4 trillion cigarettes, accounting for 40 percent of the world's total. The tobacco industry in China reportedly pays the equivalent of $141.9 billion a year in taxes. China has more than 300 million smokers, nearly a third of the world's total. More than half of adult men are current smokers. In China, smoking cigarettes by the young is considered to be a sign of adulthood. China National Tobacco Corporation is the world's largest cigarette producer and retailer in the market.

In May 2021, the World Health Organization warned that more than 1 million people had died from smoking-related diseases in China every year, or more than 2,700 deaths per day. If current smoking trends continue, this number is estimated to double by 2030. Smoking can cause many diseases, including heart disease, stroke, lung disease, and diabetes, as well as various cancers of the mouth, nose, throat, voice box, esophagus, blood cells, liver, stomach, kidney, and pancreas. It also damages the eyes, which can lead to vision loss, cataracts, and macular degeneration. Smoking can result in miseries to families, impoverished communities, and damage to the economies.

The costs of treating patients with related diseases caused by smoking are phenomenal due to the large number of smokers. With smoking rates so high among Chinese men, death due to smoking disables the main source of income for many Chinese families, hurting the poor family the most. If nothing is done to stop these trends, many millions of people in China will die from smoking-related diseases. This will reduce economic productivity and push tens of millions into poverty. It will also put more demands on China's social welfare and medical services.

Death due to smoking results in loss of lives and waste of healthcare resources. Hence, smoking must be stopped. According to research, medical expenses attributed to smoking in China are substantial, with estimates placing the total cost of smoking-related illnesses between 57-368 billion yuan per year. This represents a significant portion of the total healthcare expenditure in the country, estimated to be around 8%.

The central government should draw lessons from the Opium War, which led to the fall of the Qing Dynasty and the invasion of China by eight foreign countries. It is time for the central government to declare war on cigarettes. A slew of comprehensive draconian laws are to be introduced to curb cigarette sales and smoking prohibited in public places and inside buildings. China is introducing a total ban on cigarette advertising, promotion and sponsorship, a tripling of cigarette sales tax. The Ministry of Health is to educate school children about the dangers of smoking. Banners are placed in public places to urge smokers to quit smoking. All state TV stations will advise viewers of the danger of smoking.

Each smoker is required to apply for a permit card to purchase cigarettes in the country. Only registered companies can sell cigarettes, and they must record the number of cigarettes sold to the registered smokers (including foreigners) . This information will be transmitted to each citizen's national Healthcare Records Center. Each smoker can only buy 5 cigarettes per day, and only one member of each family can apply for a permit card. According to the law, cigarettes cannot be passed to others. A smoker has to pay hospital charges fully if his treatment is related to smoking. The law prohibits foreigners from giving or selling cigarettes to others. Foreigners who break the law will be heavily fined. Smokers have to pay higher health insurance costs.

Passengers carrying cigarettes from abroad must declare to the Customs the quantity of cigarettes purchased from abroad. Only registered smokers can buy cigarettes overseas. Passengers have to pay three times the import duty. The number of cigarettes brought from overseas cannot exceed 200. Non-smokers are not allowed to buy cigarettes from overseas. Foreigners entering the country with cigarettes must fill out a form and apply for a permit to be allowed to import them, and cannot resell or give to others as gifts. Smokers can be fined up to $1,000 if caught smoking in prohibited places, and up to $10,000 for repeat offenses. Tobacco companies are not allowed to give away or sell cigarettes to their employees.

17.9 LOW STANDARD OF MEDICAL SERVICES IN RURAL AREAS

Hundreds of millions of elderly patients in rural areas, suffering from multiple chronic diseases after years of hard work in the rice fields and inclement weather, do not receive similar healthcare services as provided to the urban residents in the cities. Many village doctors are not well-trained, and they lack medical equipment, instruments, chemical laboratories and X-ray facilities. The doctors do not have adequate supply of medicines to treat various illnesses and diseases. Of course, the central government cannot afford to build a hospital to cater to a small population in a village. In the past, the villages were normally served by nurses, who would decide whether the patient should seek help from the doctors in the city hospital. In addition , the rural residents cannot afford to buy insurance as their income is too low, especially during periods of drought or floods that affect their crops in the fields.

Village government's budgets for healthcare are too meager even to afford to build an outpatient clinic to provide emergency services to the 600 million rural residents. According to a report, the total cost of healthcare in 2024 reached 2.03 trillion yuan, accounting for 6.31% of GDP. Most of the national healthcare budget is spent on the 800 million urban residents. To improve the healthcare service in the future, the government can provide mobile medical service on a regular basis by doctors flying drones from the town hospitals to visit the various villages under their social service responsibility. The village patients will need to download hospital admission forms describing their illness so that the doctors can prepare appropriate medicines for them.

17.10 HIGH-TECH MEDICAL CARE

In the context of rapid development of AI, upgrading of the medical industry has become the focus of great concern in all countries. According to reports, in the field of medical robots in China, surgical robots account for 37%. In recent years, a number of policies have been issued to encourage the innovative application of medical robots. With the help of surgical robots, doctors are less tired during surgery and patients' hospital recovery time is shortened.

Surgical robots are mainly used in the departments of bone surgery, neurosurgery, laparoscopic surgery and vascular interventional therapy. From the patient's point of view, the operation is more accurate, the postoperative recovery is faster, and the healing is good. Minimally invasive surgery results in reduced postoperative pain, shorter hospital stay, and reduced blood loss. For doctors, the surgical robot increases the angle of vision and reduces hand tremor. The machine is smaller than the human hand and can work in a confined and narrow space, so that the doctor can work in a relaxed working environment, reduce fatigue and concentrate more. It also reduces the number of surgical personnel.

High-tech remote minimally invasive interventional surgery is popular. A number of hospitals are remotely guided to perform minimally invasive interventional surgery through the 5G technology platform. According to reports, at present, the incidence of liver cancer in China accounts for about half of the world, and about 85% of patients have lost the chance of radical treatment such as surgery and liver transplantation when they are found with liver cancer. Minimally invasive interventional therapy for liver cancer is the main treatment for these patients, which does little damage to patients and surgery can be repeated.

Based on the rapid development of China's future medical technology, the government's large investment in medical technology, and the people's high acceptance of new medical technology, telemedicine will be promoted extensively. Telemedicine includes telemedicine consultation, telemedicine education and healthcare consultation system. Telemedicine consultation creates a new connection between medical specialists and patients, allowing patients to receive consultations , treatment, surgery, and care in the local hospital, under the guidance of specialists from home or abroad, thus saving both doctors and patients money and time, and potentially saving a life in an emergency.

According to the Sun Yat-sen University Cancer Prevention and Control Center, the facility utilizes the high-throughput characteristics of 5G. This enables interventional experts to receive a patient's pre-, intra-, and postoperative CT scan images from other hospitals in seconds, allowing them to quickly plan the injection and set the ablation conditions for real-time guidance of the on-site doctors on the ablation needle approach and layout. At the same time, interventional experts can also monitor the ablation process in real time to reduce the incidence of ablation complications. The "5G+AI" remote guidance mode no longer requires the center's doctors to travel to and from local hospitals many times, saving transit time and high-quality medical resources.

17.11 CLOUD STORAGE TO STORE PATIENT MEDICAL RECORD

The government has different agencies that can record the information of patients' medical records in each province on a blockchain, and no one can change it. All hospital doctors can see the patient's information on the blockchain, and no one can falsify the information. The blockchain becomes a shared national database of patient medical records. Each doctor can create a new block on the blockchain to record the patient's information, and the other doctors can see his patient's record, but only he can change it.

The National Research Center cloud computer stores countless kinds of electronic medical records (EMR) which are digital versions of patients' paper charts, containing all their health information. For healthcare providers accessing the EMR system, a user-friendly interface is required to allow quick retrieval of patient records, efficient data entry, and integration with other tools used in healthcare settings. Centralized storage of EMRs can lead to better patient prognosis of illness, as any authorized healthcare provider can access a patient's history. It facilitates medical research by providing large datasets. Public health initiatives can be informed by real-time data analysis. AI chatbot using the patient's historical medical record and current patient medical symptoms can immediately suggest a diagnosis for doctor's evaluation and acceptance. The government has been pushing for digital healthcare initiatives.

The National Ministry of Health Medical service will use the latest IT and digital technology to innovate medical services, integrate the information resources of hospitals, and build a service platform on cloud storage. Access to the hospital outpatient and inpatient is via online registration. Patients provide illness information on the hospital admission form, using AI chatbot to process the information so that the hospital can send a message to the patient with detailed steps to go to a hospital for medical treatment. The hospital will determine the needs of the patient in advance based on the patient's personal medical records and current illness symptoms. Patients with minor illnesses may only need to go to the community's family clinic.

With AI and digital technology, hospital administration will be improved, streamlined, more cost-effective and efficient, and many current problems faced by doctors, nurses, and patients will be resolved. Pressure on doctors will be reduced. Costs of overall healthcare should be reduced. The working environment in the hospital should be more congenial. Even hospitals have digital facilities for payment using WeChat or Alipay, medical reports print out and waiting time for patients.

The rapid integration of AI into healthcare in China is transforming clinical decision-making and hospital operation. DeepSeek is now widely deployed in China's hospitals nationwide. AI helps enhance diagnostic accuracy, streamlining workflow and improving doctor-patient relations. AI-powered pathology, imaging analysis and clinical decision support systems have helped optimizing medical processes and reducing the cognitive burden on doctors. AI will integrate multimodal data sources like genomics and radiomics leading to better personalized treatment. Several universities are partnering with technology firms like Huawei to develop AI infrastructure and deploy DeepSeek on high-performance computing platforms to enhance AI-driven medical education and research.

17.12 CHINA MEDICAL TOURISM INDUSTRY

China's medical tourism industry is booming, as many Chinese large city hospitals have attracted over a million foreign patients a year for different kinds of medical treatment. Patients come from Singapore, Malaysia, Philippines, and even from some European countries. According to a report, in 2022, China medical tourism had a market worth $8.9 billion. It could go up to $46 billion by 2033. Here are some reasons that foreigners seek medical treatment in China:

- Lower Treatment Costs: Some medical treatment procedures e.g., surgeries, IVF are cheaper in China than in many European countries. The premium services in the Chinese private hospitals often cost less than equivalent care in developed nations.

- Traditional Chinese Medicine (TCM) Therapies: A significant TCM draw for European and other national patients include acupuncture, herbal therapies, and wellness programs for chronic pain, stress, or autoimmune conditions.

- Cutting-Edge Facilities: Major Chinese cities boast hospitals with state-of-the-art equipment, e.g., robotic surgery, proton therapy. China is a hub for certain procedures, such as stem cell therapy, organ transplants, and cancer treatments.

- Quick Access to Medical Care: Elective surgeries or diagnostics often have minimal waiting times compared to countries with public healthcare bottlenecks in countries like Canada, UK, and US. With visa free entry, many foreigners find entering China for medical treatment convenient.

- Proximity to Asian Neighbors: Patients from Southeast Asia, Russia, or the Middle East find China more accessible than distant Western destinations.

Many foreign patients seek treatments involving both western medicines and Traditional Chinese Medicines. Worldwide, there are 10 million people with Parkinson's disease (PD). Patients with PD can go to Hainan Province that has new experimental drugs and treatment not yet available in other countries. In the past 5 years , biotech companies conducted 750 clinical trials for PD in Hainan. There are 3 primary treatments for PD: gene therapy, deep brain simulation (DBS) and stem cell research. In stem cell PD treatment, it involves nanoparticle injection directly into the brain that could reverse the worst symptom of PD without surgery. It can cost hundreds of thousands of dollars in the US to have such treatment if the treatment is approved.

CHAPTER 18: CHINESE EDUCATION SYSTEM
18.1 PRESENT STATUS OF EDUCATION SYSTEM

The word education in Chinese language has two meanings : "to teach" and "to educate". "To teach" means to impart knowledge to the students and perpetuate the culture of the country. "To educate" refers to cultivating, guiding and training students to search for and innovate new ideas and things. Every child is like a rough diamond which needs to be polished to achieve a perfect flash to be of a high-value product. Every child has a hidden potential, but it must be nurtured through good and proper educating in order to reach his maximum potential.

A smartphone is the culmination of years of global research, integrating knowledge from different people and countries. This complex product draws on advancements in fields such as chemistry, metallurgy, electronic and electrical engineering, radio frequency engineering, material science, light sensing, software, hardware, wireless technology, satellite GPS, battery technology, biometric scanning, facial recognition algorithms, and AI chips. Does any student from high school or university understand the complexity and amount of knowledge involved in building a smartphone? Does any student understand what is involved in building a commercial aircraft? These two products help people to enjoy life in a progressive environment. This world is built with shared knowledge of people from different times and countries.

School education is a success when the students after graduation can contribute to the society and support the economic progress of the country. The government should recognize that education is a good investment rather than an expense. If the government spends $20,000-$30,000 on 10–12 years of education for each student, in the next 30–40 years, each student could make an economic contribution of $2-10 million to the society during his working life. The state gains from its investment in education. Hence, investment in education is to propel the economic progress of the country.

School education should not teach students just to pass examinations. The school should not encourage students to spend time and money to attend private tuition just for the sake of passing an examination with high marks. When the students enter the job market, their job contribution is judged by their performance or value added contribution, and not by his scholastic results. The government does not hesitate to invest tens of billions of dollars to build a highway and high speed rails network to accelerate economic development for the country.

Presently, school teaching focuses on teaching the fixed curriculum to the students in primary and secondary school. The students are expected to regurgitate the same answers given by the teachers for the questions during examination. Hence, the students do not attempt to answer the same question with an alternate answer even though their answer is correct but will be rejected by the teachers. Students are given a lot of homework to do, and the parents are supposed to supervise them at home. Many parents complain that their children are very tired at the end of the day. Their children have no time to relax or explore new knowledge beyond the text books, and feel a great deal of pressure going to school. Some parents send their children to private tuition in order that their children can score high marks during school examinations. Private tuition increases the financial burden on the parents.

This kind of education system in China is similar to a factory production line. Each year, the school produces a similar product of students. Whether these products satisfy the social requirements is of no concern to the school principal or the education department. It is up to the individual graduate student to find employment of their choice. There is always a mismatch of demand and supply of school graduates for various industrial and service sectors because of the ever-changing social and economic environment. Such a constant environment change is unknown to students. China's industrial and social revolution from 1978 to 2024 have changed dramatically through different stages of economic development, and hence there will be changing demand for graduates of various skills and knowledge.

The school education system has to keep abreast with societal demands. In the past, very few students wanted to take up chemistry subjects in the university as they felt there was no job market for chemistry graduates. However, chemistry graduates now play a significant role in the development of batteries for the EV industry. China EVs are so popular in the world market because of the leading edge design of the batteries produced by CATL and BYD.

The quality of education depends on the way teachers teach and the degree of their training, as well as their enthusiastic attitude. Some teachers lack regular and continuous training. Many teachers have different standards of teaching. The teaching quality of individual teachers varies greatly. Teachers do not encourage students to ask questions. Students' learning interests do not develop freely under the present rigid curriculum. Some schools are underfunded and classrooms poorly equipped, especially in small towns. Many classrooms are too crowded in remote towns and villages. Teachers are too tired and spend too much time doing administrative work.

18.2 Problems of Admission to Good Schools

The yearly problems of kindergarten, primary and secondary school admittance involves both public and private schools. Under normal circumstances, good reputation schools have better teachers and facilities. Therefore, it is natural that every parent wants to enroll their children in these good schools. But these schools have a limited number of vacancies. Therefore, how to accept all the student applications has become a problem for the school principal. People with connections and power have priority to be admitted, as do parents who pay an agent to gain school admittance to the school. Wealthy parents have no problems enrolling their children in good schools. They also have the means to buy property near the school to avoid driving their children to school during rush hours on the road in the morning.

Some parents feel that in China the high pressure school learning is not conducive to their children's upbringing and decide to send their children to international schools in other countries such as Thailand, Malaysia, and Singapore. In the international schools, the children have freedom to select subjects they like to study. The students are encouraged to participate in sports, and they finish their school work in school. The teachers encourage the students to ask questions. The wrong answer does not get rebut from the teacher. The children have the opportunity to learn and practice their English in school, as many students who study at the international school come from different countries and their common language of communication is English. Many parents are very happy when their children begin to converse in English with their school friends. Their children enjoy going to school every day. Although in China, the students learn the English language, they have no opportunity to speak English.

One accompanying mother in Malaysia said her son was an introvert in primary school in China, and no classmates wanted to play with him. After six months in the Malaysian International School, her son could even speak English with his classmates and played with them. Her son did all his homework at school without having to hire a tutor. Teachers adopt the American teaching system, encouraging students to ask questions, and they report to the parents on their children's interests, learning abilities and progress. The child does not face any pressure of learning in Malaysian International School. To the students , learning in Malaysian International school is fun and stress-free, and they become more sociable. The children can enter the local or overseas university after finishing high school education. Studying in a foreign country opens up the child's horizons.

18.3 EDUCATION SYSTEM UPGRADE

China is introducing AI education in primary and secondary schools. (Note: Donald Trump appointed Linda McMahon as the Secretary of Education in 2025. She is a novice in education. In one conference on AI education, she announces the US government is happy to introduce "A One" instead of artificial intelligence education in primary school) . Students use AI to do their homework, AI is used in class teaching and practical application making robot models. AI is being integrated into primary and secondary education in China, reflecting the government's strategic emphasis on technological advancement. The following projects are in the processes of being carried out:

- Next Generation AI Development Plan :

- This national strategy underscores AI education, aiming to cultivate talent from an early age. The Ministry of Education introduces AI-related content into national curricula, including coding and data science, starting as early as elementary school.

- Pilot Programs and Experimental Schools:

- Leading cities like Beijing, Shanghai, and Shenzhen have launched pilot AI programs in select schools. These often include specialized courses, smart classrooms, and partnerships with tech firms (e.g., Tencent, Alibaba) to provide tools and platforms.

- Textbooks and Resources:

- AI textbooks have been developed for K-12 students. For example, elementary students learn basics like facial recognition, while secondary students explore machine learning and Python programming. Provinces such as Zhejiang and Jiangsu have adopted these materials.

- Extracurricular Activities:

- Robotics clubs, AI competitions (e.g., National Youth AI Innovation Challenge) are held to encourage fostering hands-on experience. Companies like SenseTime and DJI often sponsor these events.

- Teacher Training:

- Programs like the "AI Teacher Training Initiative" aim to equip educators with necessary skills. Workshops and online platforms are used to address the shortage of qualified AI teachers.

- AI Tools in Education:

- Adaptive learning platforms and AI-driven grading systems are increasingly used to personalize education and reduce administrative burdens, even in rural areas through government-backed digital equity efforts.

- Curriculum Standardization:

- Implementation varies, with some regions integrating AI into existing subjects (e.g., math) rather than standalone courses.

- Conclusion: China's approach to AI in education is systematic and ambitious, driven by national policy and urban pilot programs. While progress is notable, challenges in equity, teacher readiness, and ethical frameworks remain. Compared to other countries,

China's strategy is more centralized and scale-oriented, aligning with its broader goals of becoming a global AI leader by 2030.

The prosperity of a country depends on the creative power of its people, and this power comes from knowledge, and this knowledge is acquired by themselves with the help of their teachers and parents. The speed of education is the central barometer of human progress, and the faster the speed, the faster the progress. Education should progress in abreast with time. Education should maximize the potential of all children so they can excel when they enter the job market after graduation. Each student's potential and interests vary from person to person. A fixed school curriculum for teaching is not ideal to develop each student's potential.

China in the past 10–15 years have advanced greatly in developing the high-tech industries, high speed rail, subway trains, industrial drones, aerospace, AI technology in EVs and autonomous driving, self-driving and self-parking, electric scooter, electric flying car (eVTOL) , autonomous self-driving mining truck, industrial robots, 5G communication, smart military weapons, nuclear battery, digital technology. However, China still lacks precision engineering and instruments which are prevalent in Japan and Germany.

Science and technology is advancing rapidly around the world. Many products have a short life span and become obsolete quickly. In this world, every system is a work–in- progress. In the future, the curriculum of the school is not devised for examination purposes but based on understanding natural principles and laws and cultivating students' creative thinking ability. Future courses for students rely on artificial intelligence to analyze students' interests and potential, and create independent courses for them to be educated. The current world order is in the form of big data and AI, with IoT affecting how we live and work. This is how the world that has functioned in the past 4 decades with a global supply chain and multipolar trade system connecting every part of the world is evolving constantly. China must adapt to the ever-changing world. China's education system must also adapt to the new world order. The new education system in China is technology centered and not examination result oriented.

The present school curriculum varies from province to province. Many school teachers are involved in setting the questions for school examinations. Such an education system is archaic and is not productive, and is entirely unsuitable for the present context of world events and progression. Many foreign manufacturers set up factories in China to produce products to sell around the world for the reason of cheap labor costs. Many of the products are of low value and technology. Chinese workers get very low salaries.

As the economy progresses rapidly, industrial productions of higher value products require skill upgrade and workers salary progressively increases until the labor costs begin to exceed those in other countries like Vietnam and Indonesia. Local factories begin to increase productivity by installing robots and reduction in manual workers to remain cost competitive. As China's economy is entering a multifaceted phase, with products of different grades and prices serving various countries in the world, Chinese industrial factories must adapt to the requirements of different markets around the world. When the AI powered industrial world becomes more robotic, many jobs will become extinct. There will be no jobs for welders, seamstresses, chefs, taxi drivers, cashiers, farm hands etc. in the future.

The Chinese Education Department begins to revamp the entire education system in order to meet the requirements of future generations in this competitive world and to compel China's economy to the next higher level to overtake the US economy. The Education Department prepares a curriculum using AI technology and is based on each province's population, its natural resources and local industries, for example, Inner Mongolia and Shanxi producing most coal, Inner Mongolia, Jiangxi, and Sichuan producing rare earth minerals. Hence, the school syllabus will emphasize minerals, earth science, geology, chemistry, mechanical processes, robotic application, water resources and AI technology. The students should study subjects suitable for employment to meet the local demand. There should also be vocational colleges to train students to be technicians for the industries. Skilled technicians can earn higher salaries than university graduates working in administration jobs. There is a tendency for some parents to force their children to enroll in the university even though their children are not interested in academic study.

School syllabus for provinces in Northeast plains for producing grains, North China Plain for wheat and corn, Yangtze Basin for rice will emphasize on vegetations, earth science, genetics, nutrients, environmental science, climate pattern, anatomy of plants, photosynthesis, fertilizers, plant diseases, insecticides , AI technology and robotic farming machines. School syllabus for students living in coastal regions and inland fishery industries will include subjects on meteorology, oceanography, biodiversity, water pressure and temperature of water in the sea and fishing ponds, marine ecosystems, microorganism, water pollution, fishing methods and marine lives, fish infection and diseases treatment and fish fry breeding and AI technology. Every school will be equipped with AI teaching materials in real time.

Many prestigious universities are located in the eastern part of China, and high school students from all over the country tend to apply to study in these prestigious universities. They also tend to stay in the eastern cities for employment after graduation. There are no prestigious universities in other parts of China. Hence, economic development in other parts of China lags far behind the eastern cities. Such a situation is not conducive to the balanced development of the country. Therefore, the government should establish new universities in the western part of China to dovetail to the revamped education policy, in which each province will emphasize the development of the local industries utilizing the local natural resources. After graduating from the new universities, the graduates will remain in the province to work in the local industries. Such new education development will accelerate the national economy's progress. Local industries will offer scholarships to promising students in their study. During vacations, these students will work in the industries to gain practical experience.

18.4 Education Expenditure in China

According to reports, the national fiscal expenditure on education in 2024 is 5.3 trillion yuan, accounting for 4% of the GDP. As the world's second-largest economy, China's education budget is low compared to other advanced countries. Education spending is critical for human capital development, innovation, and long-term economic growth. China's relatively low investment in education is puzzling, because the return on investment in education is astronomical! China's lavish spending on high-speed rail and airports is a far cry from the money spent on education. A large part of the national education budget is spent on first-tier cities such as Beijing, Shanghai, Guangzhou and Tianjin, where there are already many good universities and primary and secondary schools. Only a fraction of the education budget is allocated to rural areas and towns.

China should revamp the entire education system to keep abreast with time and world events. China needs to climb the ladder to be the strongest economy in the world with 1.4 billion people within the next decade. When education system reform is in line with the world order and encourages bright students to remain in China to study in local universities instead of studying in the US, China can save billions of dollars annually and this amount of money will recirculate within the country.

18.5 USA RECEIVED TALENTED CHINESE GRADUATES FOR FREE

China spends a lot of resources to nurture a lot of talented students and then let them migrate to the US for free. Many thousands of Chinese scholars after graduating from universities remain in the US to become American citizens, They mainly work in high-tech companies or do research in universities. Their research is helpful to US new weapon design that is used to improve American military power against China. In the 2023-2024 academic year, there were 277398 Chinese students in the US. They spend annually several billions of dollars in the US to support US education industries. It is time China do something to prevent brain and financial drains to the US. For example , their passport should only be valid for the duration of their studies. Extension of expiry date can be done only after one year of returning home. Here are two samples of Chinese smart scholars who remain in the US after graduation:

- Gao Huajian. He was born in Chengdu in 1963. He, at the age of 14, was admitted to Xi'an Jiaotong University to study engineering mechanics. After graduating from university, he received a scholarship to further his study at Harvard University. Currently, he is a leading expert in the field of solid mechanics and his research includes nanomechanics, nanostructure and low dimensional material mechanics, fracture mechanics, thin film mechanics in engineering and biological systems mechanics, nanomedicine, etc. His work has practical applications in aerospace, electronics, and medicine, such as improving drug delivery systems and designing stronger lightweight materials.

At the age of 24, he became an assistant professor at Stanford University. He is also the editor-in-chief and editorial board member of many international academic journals. Unfortunately, Gao Huajian has remained in the US to become an American citizen. China paid money to cultivate a highly talented scholar and gave him to the US for free.

- He Biyu was born in 1985 in Xinxiang City, Henan Province. In 2000, she gained national attention in China by entering the elite China University of Science and Technology (USTC) at the age of 14 through a special program for gifted youth. She graduated in 2004 with a bachelor's degree in biology. She went to the US to further her study and earned a PhD in neuroscience from Washington University in St. Louis, and later conducted postdoctoral research at the National Institutes of Health (NIH). Her research focuses on neuronal communication and synaptic plasticity, particularly in understanding how brain circuits develop and function. She joined the Department of Neurology at New York University School of Medicine as an assistant professor in 2016. She is now a US citizen

Michael Loong

18.6 China's Education System Needs Reform

Brain drain, especially the loss of top talent, is a problem that China needs to solve urgently. Qian Xuesen (1911–2009), the pioneer of China space exploration, once said that "science knows no borders, but scientists have national boundaries." No matter how hard and difficult it was during his early days , countless scientists after graduation in the US, tried every means to return home and help the scientific development of China. China was poor at that time. Huawei founder Ren Zhengfei also said: "In the field of computer chip manufacture, what we lack is not equipment, but talent!" Huawei has more than 80000-90000 R&D engineers, but a lot of them are foreigners, not that Huawei does not want to recruit domestic talents, but that there are too few domestic talents in mathematics and physics.

Yang Zhenning born on October 1, 1922 , after graduating from National Southwestern Associated University (Kunming), went to the University of Chicago for his PhD study. He received the Nobel Prize in Physics in 1957. He returned to China in 2003 joining Tsinghua University in Beijing as a professor emeritus. His later years focused on mentoring young physicists and promoting science education in China. He pointed out: "A scientist must have creative intuition. Without imagination, one cannot make breakthroughs" and "The most important thing in research is to always question assumptions."

The present Chinese education system concentrates on teaching students to score high marks in exams as a sign of achievement in education. The traditional goal of education in schools is to teach students to pass the Gaokao exams (college entrance exam) with flying colors so that they have a better chance of being admitted to a prestigious university. Chinese schools sometimes design exams with challenging questions to fish out top performers to prepare students to pass the Gaokao and enter prestigious universities. Critics argue excessive difficult questions can promote rote learning and stress.

Many parents feel that such an education system does not bore well for bringing up their children who are asked to do excessive homework till late at night. Extended school days, heavy homework loads, and supplementary tutoring leave minimal time for relaxation or extracurricular activities for children. Chronic sleep deprivation is common, contributing to physical and mental fatigue and stress. It seems it is a collective pressure system on students to pass Gaokao exams with high marks as the only means of achieving success in life. Not all successful Chinese entrepreneurs are graduates of prestigious universities. About 42% of the billionaires in China did not attend university.

The new revamped education system is to teach students in different provinces with syllabus emphasizing on their natural resources and related industries. Their Gaokao exam questions are more on practical knowledge rather than their theories. The new education system is a complete revamping to keep abreast with time and AI-powered technology. It is time China phases out high-pressure Gaokao exams and rote learning system, and moves towards more personalized and holistic development, with a focus on fostering critical thinking and creativity and to discourage students from attending private tuition lessons.

Chapter 19: Transportation Systems in China
19.1 Transportation Situation in China

China is a huge country, covering an area of 9.597 million km². There was no expressway or highway built across the country during the 1960s, as the country's economy was in a frail state. After the country introduced an open market economy in 1978, the government invested a great deal in building highways across the nation. "To be rich, build roads first" has been the government slogan. Highway throughout the country is a necessity to transport people and goods in order to increase economic activities for the society and industries. The speed of national progress is directly proportional to the speed of economic activities. The modern transportation infrastructure system carries not only the flow of people and logistics, but also the flow of information and capital. The system is also meant for enhancing national security during military deployment. It also helps the economy of the less developed regions by transporting their produce to the matured markets in other regions.

Expressways across the great expanse of China between the eastern and the western regions through deserts and farmlands facilitate rapid movement of goods, reducing logistics costs and enhancing supply chain reliability. This supports domestic trade and integration into global markets, thereby reducing the economic disparity between different regions. Previously, produce from the poorer western regions could not be transported to the eastern regions for marketing due to lack of expressways or highways.

Throughout the country, rural access roads have been built to connect to the highway or expressway in order to increase the living standard of the rural farmers with tourism and staycation businesses, thereby elevating the income of the rural farmers. Local tourism in the rural areas attracts self-drive families for family outings at short distances. With the network of highways and expressways, products of the cottage industries in small towns can reach nationwide and be exported. Constructions of highways and expressways generate employment.

Another important transportation infrastructure for China is to construct a network of high-speed rail (HSR) systems. China's first high-speed train line opened on August 1, 2008, connecting Beijing and Tianjin. It was built to meet the growing demand for fast and efficient transportation between the two big cities. The HSR from Beijing to Tianjin takes about 30 minutes to travel a distance of 117 km. China's HSR network has expanded rapidly since 2008. By the end of 2024, China's HSR network will be 48,000 kilometers long, making it the world's largest HSR network. China's HSR can travel up to 350 km/h (220 mph). The reason that China's HSR system can be built so quickly is due to the intense government support, construction capacity, modular design and standardization and constant innovation. The good safety record of the HSR operation has been due to the use of AI in managing and maintaining the HSR.

Chinese HSR is exported to foreign countries. In 2023, China completed and officially launched the Jakarta-Bandung HSR line in Indonesia, marking the first high-speed rail line in Southeast Asia. This project is performed under the China's Belt and Road Initiative. The HSR line, a distance of about 142 kilometers, reduces travelling time from 3 hours to 40 minutes. The projected first year ridership was 2-3 million passengers.

Before the advent of the HSR, many migrant workers working in the city rode their motorbikes long distances in harsh weather conditions of rain and snow to return home for Chinese New Year. Their bikes were fully loaded with their belongings and gifts. This happened in the early 2000s. Their arduous long journey in the cold weather conditions might take several days to reach home with many rest stops. After the New Year celebration, they rode their bikes back to their work place in the city under similar inclement weather. Now the migrant workers take the HSR during the New Year period to reach home on the same day in comfort. The Chinese government does take good care of all the citizens. The HSR fare is affordable, and the government runs the HSR system at a loss annually.

Before China introduced an open market economy in 1978 , Chinese people depended on bicycles and buses for their daily travel. There were hardly any cars on the road. The government did not have enough foreign exchange to import cars. However, China did produce trucks to transport goods. Manufacturing capability was limited to low technical skill and quality. However, after the policy of open market economy was introduced in 1978, China's economy began to flourish. Many people had money to purchase locally produced cars.

Here is a list of foreign carmakers making passenger cars in China

- Volkswagen in 1984 JV with SAIC to produce Santana
- General Motor in 1997 JV with SAIC to produce Buick
- Honda in 1998 JV with Guangzhou GAC to produce Accord
- Hyundai in 2002 JV with BAIC to produce Elantra
- Peugeot in 2002 JV with Dongfeng Auto to produce Peugeot 307
- Audi in 1988 JV with FAW to produce Audi 100
- Toyota in 2003 JV with FAW to produce Vios

- Nissan in 2003 JV with Dongfeng Motor to produce Sunny
- BMW in 2003 JV with Brilliance Auto to produce BMW 3 Series
- Ford in 2003 JV with Changan Auto to produce Fiesta
- Mercedes-Benz in 2005 JV with Beijing BAIC to produce MB E Class
- Tesla in 2019 produces Tesla Model 3 without JV

As China's economy was on the fast track with rising GDP and China became the world factory manufacturing all kinds of products at competitive prices, people's income soared. In 2009, vehicle sales surged to 13.6 million units and China became the largest market for passenger cars, surpassing the United States for the first time. Until 2020, General Motor sold more cars in China than in the US. China dominates electric two-wheeler electric motorcycle production, accounting for over 90% of global output. This includes e-scooters and e-motorcycles.

According to government statistics, the number of motor vehicles in China will reach 435 million in 2023. The number of new energy vehicles (EV) reached 31 million in 2024. In 2024, China overtakes Japan to become the world's largest automobile exporter and the world's largest automobile producer. Chinese EVs have no foreign competitors. They have expanded their production globally, establishing manufacturing facilities in multiple countries as follows:

- Thailand: BYD, SAIC Motor (MG), Great Wall Motor (GWM).
- Indonesia: SAIC Motor (MG)
- India: SAIC Motor (MG)
- Uzbekistan: BYD JV with UzAuto
- Pakistan: SAIC Motor, JV with local companies
- Hungary: BYD, Great Wall Motor

- UK: SAIC Motor (MG)
- Spain: Chery Automotive
- Belgium: Geely (Volvo/Pole)
- Brazil: BYD, Great Wall Motor
- Argentina: Chery Automotive
- Egypt: Dongfeng Motor, JV with local firms
- Morocco: BYD

Presently China produces the most electric buses in the world, leading globally due to robust government support, strong manufacturing capability, high domestic demand, and extensive export networks. In 2024, it exported over 15,000 electric buses to countries in Europe, Latin America and Asia. China's well-established battery supply chain enables cost-effective, large-scale manufacturing of electric buses. Chinese cities like Shenzhen have transitioned to fully electric bus fleets a few years ago. Companies like BYD, Yutong, and Zhongtong are global leaders in electric bus production. BYD has factories in the USA, Hungary, France, Brazil, India, and Canada to make electric buses. Several Chinese automotive companies have acquired or taken significant stakes in foreign car companies. Here are some examples:

- Geely (Zhejiang Geely Holding Group)has acquired a stake in Volvo Cars (Sweden), Lotus Cars (UK), London Taxi (UK) , Proton (Malaysia), Daimler AG (Germany)
- SAIC Motor: MG Rover Assets (UK)
- Dongfeng Motor Corporation: Stake in PSA Peugeot Citroën (France)
- BAIC Group: Stake in Daimler AG (Germany):
- Beiqi Foton Motor: Borgward (Germany)

With more vehicles on the roads, traffic flow in the cities is getting very congested with vehicles moving at snail speed during rush hours, and the problem is compounded by rapid urbanization. To resolve the perennial traffic congestion problems, the government implemented a modern and affordable subway system. According to a report, up to 2023, China's urban rail transit system, which includes subways, light rail, and electric buses and other metro-like systems, exceeds 10,000 kilometers in total length. This subway network spans more than 40 cities across the country and expands continuously in all the cities. Key cities contributing significantly to this length include: Shanghai with over 800 km , Beijing with approximately 800 km, Guangzhou and Shenzhen each with networks exceeding 500 km. Presently, China operates the largest metro network globally, surpassing other countries in both scale and growth rate.

China is a major global exporter of subway cabins and related rail transit equipment. Chinese manufacturers, led by state-owned CRRC Corporation Limited, have become dominant players in the global market due to their competitive pricing, advanced technology, and large-scale production capabilities. CRRC has delivered subway cars to cities like Boston, Los Angeles, Chicago, and Philadelphia. China also supplies metro trains to Portugal, Austria, the Czech Republic, Singapore, India, Thailand, Australia, Argentina, Chile, Algeria, Egypt, and Nigeria. China accounts for over 70% of the global rail equipment market.

19.2 Measures to Reduce Traffic Accident and Congestion

In order to reduce traffic accidents and congestion on the roads, the government introduces an advanced intelligent transportation system (AITS), which employs IoT and AI technology. The AITS can reveal instantly the real time traffic conditions, cause of vehicle accidents and congestion in various parts of the city every day. The AITS connects vehicles, roads, and people with the aims of improving traffic flow, safety, efficiency, environment and saving energy. For a new city, the AITS can be integrated with the entire urban city design to achieve the most comprehensive, optimal, and safest transport system, using the most scientific IoT and AI to manage the traffic system. The AITS will divert traffic to avoid any road involved with accidents or congestion. More subway trains will be added when the passenger volume increases.

Causes of traffic accidents in China:

- Speeding, drunk and reckless driving, improper overtaking, inexperience, distracted driving, sudden stopping, emotional distraction, poor health, flouting traffic laws.
- Poor lighting and road design, inadequate road signage,
- Defective car braking system
- Causes of road congestion in China:
- Population growth
- Peak hour traffic
- Economic growth — higher car ownership
- Poor road design
- Lack of public transit system

- Bottlenecks:
- Choke points like bridges, tunnels, or merges disrupt traffic flow.
- Illegal parking/stopping
- Special Events and Tourism — Events: concerts, sports games, tourist Influx
- Non-Motorized Traffic: bikes or rickshaws sharing roads with cars and trucks

Addressing traffic accidents and road congestion require holistic strategies, including infrastructure upgrades, public transit investment, smart traffic management, policies promoting alternative transportation and enforcement of traffic rules. In 2023, there were 60,028 deaths and 253,895 injuries due to traffic accidents which cost an estimated 1–3% of GDP annually, including healthcare and lost productivity. Families often bear emotional and financial burdens. As of 2024 the total vehicles registered are 453 million, of which 31 million are EVs. Traffic accidents will increase with increasing vehicle population in the future. Since almost all traffic problems are caused by individual behavior and attitude, with the help of AI and IoT, autonomous vehicles will eliminate human error problems. More electric buses and subway trains will be added to cater to the increasing population. There will be lanes separated from the main roads for cyclists and slow moving vehicles.

When designing a new city, it is crucial that the commercial center should not have too many commercial buildings concentrated in a small area. Office buildings, shopping malls, theaters, shopping centers, entertainment centers, government offices and schools should be spread over a wider area to avoid too many people rushing to the same location during rush hours. The government should restrict population growth in the large cities by encouraging new industries located in the

2_{nd} or 3 tier cities. Traffic congestion will be bound to get worse when the population keeps on increasing. More car rental companies will offer autonomous EV taxis for hire to reduce vehicle population. All the offices in the city practice staggering working hours for employees to avoid traffic congestion. To reduce daily congestion at certain road junctions, the local government will impose a congestion surcharge during peak hour periods. Such a surcharge would discourage some motorists from entering the congested area during peak hour periods.

19.3 EVs Taxi and Rental Company

China has now overtaken Japan as the largest exporter of electric vehicles. Chinese automakers could set up car assembly plants in other foreign countries to produce electric vehicles, together with a battery manufacturer to act as a supply chain. Such integrated EV manufacturing will achieve unchallenged competition in the foreign EV market. Chinese automobile manufacturing companies can produce autonomous driverless taxis for customers who are disabled and without driving license, intoxicated, for mothers with babies or young children, for people with no time to send children to school and people with special problems going shopping in heavy rain or snow. To those people who live in the cities, they do not need to buy a car for mobility as it is easy to hire an autonomous EV taxi for daily use.

It is cheaper and more convenient to hire an autonomous EV taxi without the trouble of paying tax, costs of maintenance, insurance, depreciation, and repair. Repair costs can be very high due to high labor charges in the US and Western countries. Women and elderly people would find trouble looking for a repair shop to repair their own car or changing tires, and the inconvenience involved in sending their own car to and retrieving it from the repair shop. Also, they may be scammed for paying for repairs that are not done.

With autonomous driverless taxis installed with video cameras and Wi-Fi, women or intoxicated passengers will feel safe travelling at night. Taxi companies will have their own repair shops and their cars are maintained in good working conditions for hiring. Every autonomous driverless taxi is tracked and monitored using GPS by the company's headquarters control center. A family with young children can buy a Chinese EV that can park itself in a car park and be retrieved autonomously.

With Chinese EV manufacturers setting up EV assembly plants in foreign countries, they can sign long term contracts to supply EVs to rental car companies. Rental companies have various EV models to suit various customer's requirements. Rental companies would have EVs available at every airport and train stations for short and long term rental. Emergency services are available to help customers. All Chinese EV manufacturers will have charging stations installed at strategic locations around the country. Chinese EV manufacturers think about long term investment with help from the local government. Every rental car is tracked using BeiDou in the rental company headquarter control center.

CHAPTER 20 CHINA HIGH-TECH AIRPORT
20.1 AVIATION INDUSTRY IN CHINA

The People's Republic of China was founded in 1949. At that time, China's diplomatic relations with the US and the West were not friendly. As European and American countries banned the export of aircraft to China, China could only use the Soviet-made Ilyushin and Tupolev series of passenger aircraft to fly domestically. China had no direct flights to Europe or the US, and so foreigners had to pass through Hong Kong to enter China. On February 21, 1972, US President Nixon arrived in China for an official visit. In the same year, China ordered 10 Boeing B707 aircraft. In August 1973, China received its first Boeing B707 and began international flights. On January 1, 1979, China and the US established diplomatic relations.

Since 1989, China's civil aviation industry has made rapid progress. China's aviation operations and system had undergone unprecedented changes. The advancement of civil aviation has contributed greatly to China becoming the second-largest economy in the world. Integrating aviation, high-speed rail and highway transportation had promoted national economic development. The development of China's aviation industry can be said to have kept pace with China's economic progress. According to a report, as of 2024, there are 270 civilian airports in China.

China plans to build additional 215 airports by 2035 in an effort to turn China's ambition to become a civil aviation power in a new era, in conjunction with the introduction of the Comac airplanes C909 and C919. Soon, C929 will compete with the wide-body airplanes B787 and

A350 produced by Boeing and Airbus respectively. China's demand for air travel will surpass that of the US by 2035 and account for 25 percent of the global total. According to a report, China's civil airports handled 1.459 billion passengers in 2024. In 2024, China saw a significant surge in international visitors, with the total number of inbound tourists reaching 132 million, a 61% increase year-on-year. With more countries being given visa-free travel, more foreign inbound tourists will arrive in China by air in the future.

With a huge domestic market and China's increasingly large, affluent population, China will drive the development of international aviation in the Asia-Pacific region. Driven by rapid economic growth and rising incomes in the Asia-Pacific region, more than half of the world's new passenger arrivals over the next 20 years will come from this region. In 2022, ASEAN countries have risen to become the largest trading bloc with China, with import and export trade volume reaching 6.52 trillion yuan. Therefore, the volume of passenger and cargo movement between China and ASEAN countries will grow progressively. China has concluded visa-free agreements with many European and Asian countries, so the number of international visitors to China will increase rapidly. In 2019, China's newest airport, Beijing Daxing, was officially opened. It is the largest airport in the world, and has a capacity to handle 100 million passengers a year.

To boost the local economy, the Chinese government is adding new airports in all provinces. Airport investment is profitable, as seen in many countries. It collects rents from shops and restaurants, aircraft landing and parking fees, car parking fees and many other ancillary services provided to other businesses. According to a report, there are 250–300 cities in China, each with a population of more than 500,000 people , but not every city has an airport. China needs to build new airports to boost the economy of every city. The aviation industry can help accelerate the local economy.

Many cities in the western regions have been isolated from the economic advancements in the eastern regions. The western region of China is known for its stunning natural scenery, including mountainous landscapes, deserts, high altitude lakes, lush forests, and diverse ecosystems, making them popular destinations for local and foreign travelers seeking breathtaking views. Some notable areas include the Tibetan Plateau, the Kunlun Mountains, the Tian Shan Mountains, and the Jiuzhaigou National Park.

Many old airports also need upgrading. It is expected that by 2035, the development of general aviation aircraft, short-haul transport, business travel and tourism will reach the level of developed countries. Comac and other domestic aircraft manufacturers can build small capacity short-haul airplanes to serve the smaller cities. In Europe and the US, private planes are very common, especially in the US, almost every large corporation has its own airplane as a means of transportation for their executives. They can fly to another city for a meeting in the morning and return to their head office in the afternoon. General aviation is important to China's national security.

20.2 High-tech Design of Large Chinese Airports There are 10 large international airports in China:

- Guangzhou Baiyun International Airport (CAN) : This airport had the highest passenger throughput in China, a key hub for southern China and Southeast Asia.

- Chongqing Jiangbei International Airport (CKG): This airport had the second-highest passenger throughput in China, serving Southwestern region of China

- Shenzhen Bao'an International Airport (SZX): A busy tech hub with strong domestic and international connectivity that can handle 50 million passengers annually

- Kunming Changshui International Airport (KMG): One of the busiest airports in China as a gateway hub to South Asia and Southeast Asia,

- Shanghai Pudong International Airport(PVG): A major hub for Air China, and a hub for China Eastern Airlines, Shanghai Airlines, and China Southern Airlines

- Beijing Daxing International Airport (PKX): A modern aviation international hub with a massive terminal and capacity for over 100 million passengers annually.

- Chengdu Tianfu International Airport (TFU) : This modern airport aims to handle up to 120 million domestic and international passengers annually at full capacity.

- Xi'an Xianyang Airport (XIY): A key airport in northwest China, serving the ancient capital and a growing logistics hub

- Hangzhou Xiaoshan Airport (HGH): It's one of the top 10 major domestic airports in China and a top 100 airport in the world

All the airports are designed with the latest technology and artificial intelligence. China has introduced its first artificial intelligence facial recognition technology to analyze whether flight crews are feeling anxious or have unusual facial expressions to ensure flight safety. This is to prevent any suicidal crew members from entering the cockpit. All the latest airports will apply AI video analysis technology to ground service operations, monitor any illegal activities, provide early warning systems, timely and accurate forecasts, and improve airport emergency response capabilities. The airport has a security network. The service operations center can integrate security systems to provide three-dimensional security, integrated fire protection and the establishment of automatic safe flight areas. The center will enhance security in the terminals, cargo areas, tarmac and other public areas.

According to reports, the Civil Aviation Administration of China (CAAC) and Huawei signed a strategic cooperation agreement, in which the two sides agreed to cooperate in the design of intelligent civil aviation operation, to promote scientific and technological innovation, set standards and train employees. The intelligent operation center can monitor the five major business streams of flights, passengers, baggage, cargo, and transportation in real time. The center can also optimize forecasting, early warning, collaborative operations, and intelligent decision-making. At the immigration check point, passengers use facial recognition technology as a unique means of identification to ensure the smooth flow of entry and exit. According to statistical analysis, self-service face recognition can reduce queuing time by 20%, reduce airport personnel by 50%, and improve self-service boarding efficiency by 40%.

20.3 AIRPORT DEVELOPMENT IN CHINA

World-class airport hubs in China's major economically developed regions are expected to be strengthened over the next 15 years. China will also accelerate the construction of four international cargo airport centers, including Zhengzhou and Tianjin in Henan province, Hefei in Anhui province, and Ezhou in Hubei province. Yiwu is China's international trade center and should have an international cargo airport hub. China will have about 40 airport hubs, which will not only closely connect domestic regions, but also strategic international airports. With more visa-free travel, more foreigners will visit China, thereby boosting tourism receipts to be on par with the US on a per capita basis. China's long history and many natural scenic places should be attractive to tourists from around the world.

According to the "China Standards 2035" plan, a fully functional, three-dimensional and interconnected land, sea and air transport network will be formed. The network can promote effective economic activities, reduce the cost of enterprises, and benefit economic development. By 2035, China will have about 450 new civil transport airports. This means that China will add more than 150 new airports, an average of 10 new airports per year, including cargo hubs. China's retired passenger aircraft can be converted into cargo aircraft to reduce operating costs.

As China is already the factory of the world, there is no reason China cannot become the world's largest international freight hub. Chinese online shopping has been popular in many countries, so the demand for air freight is high. China can become the world's largest center for converting used passenger airplanes to freighters as China is the world's biggest freight forwarder in the future and China will soon overtake the US as the largest economy of the world. A recent 2024 report indicates China accounts for nearly 31.6% of global manufacturing output, compared with 15.9% of the US. China trades with over 150 countries and regions and is a top trading partner with many of them. Annually, China has a trade surplus in world trade.

20.4 New Smart Airport Design

A irport is an important infrastructure in every country around the world. It encompasses human connectivity, global trade, economic activities, logistics, tourism, cultural exchange, national security, international relations, and social mobility. A country without an airport will be isolated from the rest of the world. Every day, millions of passengers, workers, taxis, buses, and freight pass through the airport's gates. The main commitment of an airport is to ensure on-time arrival and departure of flights for the passengers while maintaining a simplified and enjoyable experience for passengers, shippers, and airport operations. However, it can be fulfilled only in a security context. Security is at the heart of airport operators. Airport designers must pay attention to two important areas to ensure that the airport is well-designed to meet its goals. The two areas are:

- Siting of airport:

- The site should be suitable for building an airport that can be used for many years and has sufficient open space for future expansion. The size of the airport must be able to cope with the passenger volume in the next 30 to 50 years. The surrounding area must be free of human settlement and industrial activity. It is important to consult the Meteorological Office to check the weather conditions of the site. Severe weather conditions such as thunderstorms, heavy snow, heavy rain and extreme winds have a great impact on flight safety. The direction of the runway must not face any strong crosswind direction. The airport must have a reservoir for water supply and fire protection. Rainwater will be collected and run into a nearby reservoir. Since the airport must operate 24/7, the airport power supply is vitally important, and hence the power supply cannot be interrupted to affect the safe operation of the airport. A standby emergency power supply is needed to operate essential services in case of massive power stoppage.

- In order to ease traffic congestion on the highway to the airport from the city, a rapid transit subway system is being constructed connecting various parts of the city to the airport. This will facilitate airport workers to travel daily between their homes and the airport. There are some airlines that provide check-in facilities at certain hotels for passenger convenience. After check-in done at the hotel, the passengers do not need to rush to the airport check-in counter and can go straight to the departure lounge. Check-in operation at the hotel will help reduce the cost of airport passenger services by the airlines. Airport staff receive a higher pay due to extra expenses of transportation and higher food prices, and also incur long travelling time to and from work, especially to those who work the night shift.

- Aircraft accidents on airport ground:

The airport's apron, taxiways, and runways are to be designed with complete safety. Historically, many aircraft ground accidents have occurred on runways and taxiways. Most runway accidents are caused by pilot error, which might be due to pilot's inattentive action, inexperience, fatigue or operational errors. Air traffic controllers are responsible for assigning the runway for an airline to takeoff, flight direction and leveling off. They need to prioritize landing of an aircraft before another aircraft can take off from the same runway. The job is demanding, difficult, and complex. A small mistake can lead to a midair collision or runway incursion.

According to a study by the Flight Safety Foundation, about 96 percent of runway accidents and 80 percent of deaths caused by runway accidents are caused by aircraft deviating from or overshooting the runway. The Flight Safety Foundation study found that off-runway flights are the most common type of runway accidents. Misjudgments in speed, altitude, or wind conditions; delayed go-around decisions; improper flare techniques during landing are common pilot errors. High cockpit workload and fatigue could impair decision-making and reaction times.

According to research by the Flight Safety Foundation, the second most common type of runway accident is a runway incursion. The deadliest runway incursion in history occurred on March 27, 1977, at Los Rodeos Airport on the Canary Island of Tenerife, Spain, when a KLM Boeing 747 attempting to take off without flight clearance collided with a taxiing Pan Am 747, killing 583 people. On January 2, 2024, a Japan Airlines (JAL) Airbus A350 caught fire after colliding with a Coast Guard Bombardier Dash-8 aircraft while landing at Tokyo's Haneda Airport. Five of the Bombardier's six crew members were killed. All 379 people on board the JAL survived. On 23 November 1995, a Gulf Air A340 taxiing from London Heathrow Airport collided with a British Airways B757 at the waiting area. The wing tip of the A340 struck the tail of the B757, causing minor injuries to two of the 378 passengers on board and damaging both aircraft.

20.5 New Intelligent Aircraft Towing Operating System

China has invented an autonomous aircraft towing system. It is designed to operate using information technologies such as the IoT, cloud computing, big data, BeiDou satellite positioning and navigation, geospatial information, artificial intelligence, sensors, and 5G communication systems. The system has a towing platform that raises the nose landing gear of the aircraft and then locks into the towing position. The system uses a number of sophisticated sensors to position itself to engage the nose landing gear of the aircraft. It is activated and controlled by the control tower operator. The plane's flight number is programmed into an automated machine that tows the aircraft of any model to the designated arrival gate. Before the aircraft takes off, the same automated system will use the aircraft's departure flight number to tow the aircraft to the designated aircraft holding area for final takeoff.

The new autonomous aircraft towing operating system will eliminate ground incidence and improve the airport efficiency and safe airline operations. When the aircraft lands on the runway, it heads to the first open taxiway and onto the aircraft holding area for the automated towing aircraft to connect. During towing, all aircraft engines are shut down to eliminate carbon emissions and noise. In many busy international airports such as London, Tokyo, and New York, the normal taxiing time of the aircraft may take 15–20 minutes due to traffic jams, with multiple stops and starts to reach the terminal, resulting in increased tire wear, engine fuel consumption, and aircraft operating time. As a result, airline operating costs will increase.

A Singapore Airlines Boeing 747-412 tried to take off from the wrong runway at Taipei Airport on October 31, 2000. The plane crashed into the construction equipment on the blocked runway, killing 83 of the 179 people on board. The aircraft was destroyed. This tragedy could have been avoided if there was an automatic aircraft towing system. The accident was caused by a pilot error of entering a wrong runway.

High traffic density on the tarmac and on the taxiway is prone to incidence and accident. Aircraft collisions are a frequent occurrence. In terms of return on investment, the cost of this automated aircraft towing system is negligible compared to the annual financial loss of airport congestion and accidents and the reduced frequency of flight takeoffs and landings resulting in less efficient airport operations. With the automatic aircraft towing system, airlines can save on aircraft fuel and maintenance costs, the need to purchase aircraft towing trucks and hire drivers, reducing ground staff and human-caused incidence and increasing the efficiency of aircraft operations.

Automatic aircraft towing operation is controlled by the airport aircraft movement control center. The departure time of each aircraft is precisely controlled so that there is no shortage of terminal gates. The new automated aircraft towing operating system completely eliminates the problem of collisions between two aircraft, and between aircraft and ground vehicles or buildings. The problem of planes taking off from the wrong runway will not happen again. Each flight only has to wait one minute to line up at the holding area for takeoff. In addition, airlines will save a lot of fuel costs because the aircraft will not need to carry extra fuel from the departure terminal to the holding area. Normal waiting in line for departure and takeoff can take 15 to 20 minutes at busy airports.

Whenever there is a long delay on the tarmac due to an aircraft ground incident, it can cause severe disruption to airline flight schedules, especially international flights. Long delays can cause pilots, cabin crews and passengers to disembark and to be accommodated in hotels, resulting in additional room, food, and transportation costs. The delay will disrupt downstream flights across the network. Similarly, aircraft maintenance schedules would be affected, disrupting the engineering department's manpower plans. As a result, every prolonged flight delay has a ripple effect on the operations of the entire airline network.

20.6 STATE-OF-THE-ART TWO-LEVEL AIRPORT TARMAC DESIGN

Airport tarmac activity is a potentially hazardous work environment. Servicing, maintaining and supporting an aircraft is a round-the-clock effort that requires cargo handling personnel, refueling personnel, toilet waste and water service personnel, catering support personnel, de-icing/anti-icing workers, airport administrative staff, aircraft and equipment service personnel, maintenance personnel and many other personnel to keep aircraft operation on time. The pace of work is fast and hectic to meet airline schedules. The tarmac is crowded, noisy, and packed with a wide variety of vehicles traveling at different speeds. Airport air pollution is serious.

Due to these environmental and human factors, especially during the peak holiday season, bad weather conditions and frequent airplane delays, airport tarmac incidence frequently happens resulting in angry passengers and disruption to airline flight schedules. In order to alleviate these problems and the consequences, the future design of new airports in China would utilize innovative design and high technology to minimize ground incidence and improve safety and productivity.

The most important feature of the new eco-friendly airport design is that the airport tarmac has two levels, separating aircraft movements from other associated human activities. The ground level floor or the apron is restricted to aircraft parking and aircraft towing, passenger, aircrew, aircraft maintenance staff activities. There will be no other traffic on this ground level. The second level or underground is used for cargo loading and unloading, aircraft catering, cleaning and refueling activities. All vehicles are electric. This new two-level tarmac design will reduce or eliminate traffic congestion and incidence on traditional tarmac, as well as reduce the impact of airport noise pollution. Bad weather such as snow, rain, and strong wind does not affect underground work.

When vehicular traffic is moved to the underground level, airplanes can park closer to the terminal, speeding up disembarkation and boarding of passengers. There is no need to install large, long, expensive aerobridges at new airports. It normally takes about 10 minutes to maneuver the long aerobridge to connect the airplane to the terminal after the aircraft arrives at the terminal gate. The newly designed tiered tarmac can save 20–25 minutes of passenger boarding and deplaning time per flight. The time saved is a source of revenue for both airlines and airport authorities.

With the new design isolating aircraft activity from other vehicular traffic and a corresponding reduction in noise level, people working underground will feel less stress and have more room to move around without being affected by the harsh weather and airplane movements. In addition, a new system for removing aircraft toilet waste without the use of mobile trucks will be incorporated underground thereby saving airline costs. Similarly, there is an automated system for delivering food trolleys from the airport catering kitchen to the aircraft galley.

The goal of the new airport operation is to reduce costs for airlines and airport authorities and to provide convenience and safety for passengers and freight forwarders. Eco-friendly airports usher in a new era for the aviation industry, promoting greater productivity and efficiency. The airport's new distinctive features are to reduce carbon footprint and improve round-the-clock operation, especially in winter with snow and harsh climates. Rainwater is collected into a big pool for fire fighting and airport cleaning.

The new airport design includes a fast aircraft refueling system. The aircraft fuel is delivered from the underground storage to the aircraft refueling station on the tarmac, eliminating the need for manually operated refueling trucks, thus reducing refueling costs and traffic congestion and air pollution. The aircraft can be refueled automatically from several pumping stations at the same time to reduce aircraft downtime.

The aircraft dispatcher inputs passenger load, cargo, aircraft reserve fuel and aircraft engineering requirements into the company's aircraft flight operating system. He uses the latest computer software to calculate flight sector fuel requirements and fuel distribution in the wing tanks. Finally, before the plane even takes off, information about fuel totals and the plane's center of gravity is transmitted to the pilots, who calculate the aircraft flap position for takeoff.

20.7 ALL-WEATHER RUNWAY

Every aircraft can land easily on the new all-weather fully visible runway with the latest automatic guide slope and transponder, crosswind speed detection and control equipment. A windproof wall is installed to protect the aircraft from wind shear effects on the runway. When ice particles or snow fall and contact the runway sensors, the heating system automatically activates to keep the runway ice-free. Every year, many northern airports are blanketed in fog during winter, with severe visibility of just a few meters, causing long flight delays.

Fog is formed in winter when the ground cools rapidly at night due to clear skies, causing the nearby air to cool to its dew point and condense into tiny water droplets, creating fog. The smart airport is not affected by fog because it has an antifog system device that releases hot air into the sky near the runway to prevent fog from forming. There are instruments around the airport that measure the relative humidity of the air. When the relative humidity reaches about 80-85%, hot air is released.

In severe winter, underground pipes rise above the ground to spray hot air over the aircraft wings and fuselage, preventing ice buildup on the airplane's skin. Therefore, the aircraft does not need a de-icing truck to remove snow and ice. The economic benefit of preventing ice accumulation is greater than that of deicing. In severe cold weather, the de-icing operation can take several hours to complete, thus causing flight delays. Long delays of many flights can cause disgruntled passengers to lose their temper and create chaos for airline staff. The new design allows airlines to save on equipment , manpower and operating costs. It also improves airport operation and productivity.

20.8 REAL-TIME AUTOMATIC RUNWAY FIRE PROTECTION SYSTEM

A I will be employed to determine the optimal solution for fighting fires from crashed aircraft on the airport runway, utilizing a combination of drones, robots, and fire hydrants. From historical data of various crashed aircraft catching fire from all the airports around the world, AI bot can suggest various solutions of fighting aircraft fires on the runway. Finally, a system of different processes of how to fight various types of aircraft fire is established using water, foam, drones, robots, and fire retardants. In July 2013, an Asiana airplane flying from Seoul crashed at San Francisco airport. 16-year-old Ye Meng Yuan from China survived the crash but died as she was run over twice by a fire truck. The driver could not see her as she was completely covered by foam spread by other firefighters after she escaped from the airplane escape slide.

In an emergency evacuation involving fire, foam, fire retardant and water, the scene of the fire fighting could be frantic, chaotic and with obscured visibility during the night. In the future of airport design, fire trucks will not be used. Fire fighting will be all performed by drones, robots, and fire hydrants that can be rotated 360°and buried in the ground at strategic locations adjacent the runway. They are monitored and controlled by the fire department. Modern drones and smart robots will be widely and actively used to fight airplane fire. They are more effective than human firefighters, who are affected by the intense heat from the burning aircraft.

All new airports will be equipped with video camera devices along the runway border to detect any debris or foreign objects falling from the airplane. These debris can be detected and immediately washed away by underground water jets. Robots will then be dispatched to clear the debris. Every year, many airlines experience tire bursts during takeoff. The resulting tire debris poses a serious concern for subsequent aircraft takeoffs, as it can be ingested by an engine, leading to severe damage. Such a protection system could have prevented the July 25, 2000, Air France Concorde crash at a Paris airport that killed all 109 people on board. The accident happened when a metal strip fell off from a previous aircraft and left on the runway. It was run over by the Concorde and damaged the airplane tires, a piece of which hit the airplane wing fuel tank resulting in a rupture which led to a fuel leak and fire, The airplane was destroyed.

The new smart airport is equipped with heavy cranes to lift disabled planes on the runway. The crippled aircraft can be towed to a safe place on a giant platform so that the airport runway can resume operations as soon as possible. The giant platform is an assembled structure, joined by individual smaller platforms to create a flexible large platform that can move a paralyzed aircraft, large or small. The platform is cheaper than the usual use of expensive airplane airbags which are used to move a disabled airplane and which require periodic maintenance. This innovative aircraft transport platform equipment for moving disabled has huge cost savings for all airlines in China. China can export this innovative tool to other countries. The heavy cranes are commercially available, and the platform is owned by the airport authority and the cost is shared by all the operating airlines.

20.9 POOLING OF AIRCRAFT ENGINEERING SPARE PARTS

A t selected airports in China, all commercial airlines and aircraft components and parts suppliers establish a joint centralized spare parts warehouse to store critical spare components and parts required by the airlines. All spare parts are supplied and monitored by their respective manufacturers. When an airline needs spare parts, it must pay the manufacturer the rental of the parts. This kind of joint warehouse system is better than the old system, in which each airline has to provide a large number of spare parts at all airports nationwide and overseas stations.

It is very costly to maintain the spare parts inventory at all the stations. Problems arise when the spare parts need to be recalled for modification. In addition, some expensive spare parts may not be used for several years, thus increasing the inventory cost. This joint centralized spare parts warehouse is of great benefit to every airline as the spares can be shared by all airlines, rather than each airline storing the same inventories.

The new terminal gate in China is equipped with an automatic tire pressure inflation system and tire change equipment jointly operated by the airline and the tire suppliers, having their spare tires stored in a common warehouse. Individual airline engineers operate the system. Tire pressure check and tire inflation operation is in accordance with individual airline regulations. This system will eliminate common costly aircraft maintenance problems stemming from low tire pressure and human error causing tire ruptures and explosions.

Due to various factors, including a faulty tire gauge, a lack of employee discipline, or manpower shortages, low tire pressure may go undetected, resulting in an underinflated tire. This underinflation, combined with a long taxi distance and the subsequent flexing of the tire carcass, causes the tire temperature to rise, potentially reaching a critical level during takeoff. Tire rupture will occur, causing serious consequences. Tyre explosions have caused a number of serious aircraft accidents. Such centralized warehouses improve the efficiency of resource management to help airlines reduce costs. Today, tires are leased by airlines from suppliers and the cost of leasing is based on the number of airplane landings. It's a win-win formula for both the airline and the tire supplier.

20.10 AIRCRAFT HANDLING ACTIVITIES AFTER ARRIVAL AT TERMINAL GATE

The in-flight catering kitchen and food trolley service utilizes an underground tarmac automated delivery system to provide instant delivery of food trolleys to the aircraft galley. Upon aircraft arrival at the terminal, the system will unload the empty food trolley from the aircraft galley and return it to the catering kitchen for further action. Such automated delivery systems eliminate the need to use ground vehicles and drivers for delivery and retrieval of food trolleys. This automated electronic control system allows for quick changes in in-flight catering needs. This system contributes to significant reduction of airline operating costs, ground traffic accidents, air pollution and noise, and increasing airport productivity and efficiency. Such an automatic delivery system provides hygiene safety and less involvement of human handling of food delivery.

Currently, airlines buy their own food trolleys, which are maintained by the airline's commercial department. Food trolleys often suffer from damage and need to be repaired by a repair shop approved by the civil aviation authority. It is difficult for commercial department personnel to assess the value and quality of repairs. In addition, the trolleys require steam cleaning to remove odors and stains. Many trolleys are managed and controlled by airline commercial staff at different stations. Also, the trolleys at different stations may be damaged and require replacement from the headquarters. Unless the trolleys are managed and controlled by dedicated technical staff, It is not easy to maintain all the trolleys in good quality working conditions at every station throughout the network.

It is more cost-effective for airlines to outsource all the inflight food trolleys to a leasing company, which can manage the trolleys professionally. Because of the large scale of the business from various airlines, their unit cost of owning the food trolleys is low. As a result, airlines will save money and eliminate the hassle of managing trolleys in all stations, and staff cost is reduced accordingly. Normally, airline staff are paid a higher salary than non-airline staff. Managing the trolleys is not part of an airline business. Similarly, the flight kitchen is not part of an airline business and hence is outsourced.

20.11 Air Freight Industry

China is the world's largest exporter, so it should be natural for China to establish the world's largest air cargo company. Air freight from China can reach every customer in a foreign country within one day and is suitable for transporting goods with short expiration dates, high value goods such as computers, mobile phones, medical equipment and medicines. However, presently, China does not have the world's largest air cargo company. The following are 3 largest foreign air cargo companies in the world:

- FedEx Express, USA, with 698 cargo planes, established in 1971
- UPS, USA, with 291 cargo planes established in 1988
- DHL Aviation, Germany, with 202 cargo planes, established in 1969

These three foreign air cargo companies account for most of China's international courier service. FedEx operates more than 300 international flights per week to and from China. It has over 100 ground stations and 102 branches in China and has gateways in Beijing, Shanghai, Guangzhou, Shenzhen, Qingdao, and Xiamen. UPS has over 100 operating facilities in China and has hubs in Shanghai and Shenzhen. DHL has hubs in Shanghai (Pudong International Airport), Hong Kong, and Beijing. It employs over 22,000 staff at over 100 offices and service centers across the country. DHL employs automated warehouses with robotics and AI-driven supply chain solutions, and digital platforms for real-time tracking and customs clearance.

These three air cargo companies have been well established for many years with experienced staff working around the world. Their operations are seamless, efficient and working closely with established logistic companies around the world. Their operational language is in English and follows international aviation regulations and rules. China air cargo operation is at an infant stage. Here are 5 cargo airlines in China, and they are small in comparison to foreign air cargo airlines:

- SF Airlines, with 70 cargo planes, established in 2009
- China Postal Airlines, with 39 cargo planes, established 1997
- YTO Cargo Airlines, with 13 cargo planes, established 2015
- China Cargo Airlines, with 17 cargo planes, established 1998
- China Southern Cargo Airlines, with 17 planes, established 2001

SF Airlines is China's largest air cargo carrier. Founded in 2009, the company is headquartered in Shenzhen Baoan International Airport. The company has a fleet of 70 Boeing 737,747,757,767 freighters, making it the largest cargo airline in China by fleet size. There are many opportunities for China to capture a larger share of the global cargo airfreight market. China has become ASEAN's largest trading partner. China-ASEAN air cargo volume is expected to grow rapidly. Trade is also growing between China and India, and between China and Central and Western Asia and Russia. Therefore, China needs to expand its share of airfreight business to compete with the foreign cargo companies in the market.

According to the International Air Transport Association, China needs 650 freighters in the next 20 years. This is a golden opportunity for COMAC to produce the C919 freighter for the domestic and international market. With the rapid development of e-commerce in Asia and elsewhere, the air cargo business will grow accordingly. China should seize the opportunity to become the world's largest air freight market.

20.12 AIRLINE UNIT LOAD DEVISE (ULD) OPERATION

When an airline has an air cargo terminal at every airport around the world, it can be a tough job to maintain a high level of availability of ULD at all times to meet customer's demand at different times. During the peak seasons, domestic exporters ship several times more goods abroad than usual, and hence they need more ULD. Many empty ULDs will need to be returned to the original base to meet the next shipment requirement. Logistics operations by headquarters airline staff dealing with the return of empty ULD to original stations present difficulties as they do not deal with the shippers at various places. ULDs are damaged quite frequently during operation. The ULD baseplate is often damaged, causing the ULD baseplate to be unable to be locked in position inside the aircraft cargo hold. The affected ULD will be replaced, and a flight delay might ensue.

The airline cargo handling staff will face many daily questions like: a) how many ULD, pallets and cargo nets are scattered around in domestic and foreign airports. b) how many ULD, pallets and cargo nets are missing and how many of them are damaged beyond repair c) how many ULD, pallets and cargo nets require replacement. d) what is the turnaround time for the repaired ULD, pallets and cargo nets. These questions are not easy to answer unless the airline has a computerized system to monitor and track all the assets scattered over the entire network.

Hence, it is better for the airlines to outsource the ownership and management of ULD, pallet and cargo netting to the logistics companies as they deal with the various shippers daily in different parts of the world. Loading and unloading of the ULDs and pallets are still to be performed by the airline staff to ensure compliance with safety rules and government regulations. The airline should only be responsible for carriage of the consignment to the destination for the consignor.

Airlines cargo loaders must exercise care during loading and unloading of the ULDs and pallets in order to avoid damaging the aircraft sidewall structures and loading systems caused by improper operations. The most common damage is to the liner of an aircraft's cargo hold, caused by containers or pallets of cargo scraping along the inside of the aircraft sidewalls. The liners are made of very expensive materials with special fire resistance properties and must be replaced if they are damaged. Replacement is a time-consuming activity. The other damage that can occur is to the aircraft cargo floor power drive units. These drive units are used to move the ULDs and pallets to the correct position, and the unit rollers are also used to support the ULD and pallets. Improper loading of ULDs could damage unit rollers.

Chapter 21 Commercial Aircraft Manufacturing in China
21.1 Commercial Aviation In the 1990s

After the end of World War II, the economies of all countries except the US were severely hit. The US economy suddenly soared, opening a massive air service. The most widely used commercial aircraft was the DC3 made by Douglas Corporation. Production of the DC3 peaked worldwide, with more than 16,000 commercial and military versions. Some DC3 aircraft are still flying today. The Douglas DC-3 was a propeller aircraft that had a lasting impact on world aviation in the 1930s and 1940s and during World War II. The aircraft was simple to fly, with a cruising speed of 241 km/h, a low altitude of 2,590 meters, and a capacity of 21–30 passengers. The plane had no radar and the pilot had to fly visually. There were no night flights, no instrument landing systems, no emergency oxygen, no public address system, no air conditioning. Passengers drank hot drinks from thermos bottles. The toilet consisted of a hole at the bottom of the fuselage.

During World War II, the DC3 was called into combat, and the armed forces of many countries used the DC-3 and its military variants to transport troops, cargo, and the wounded. The main reason the DC3 is so popular is that the cabin is unpressurized, so there are no mechanical fatigue issues. The whole aircraft mechanism is very simple and therefore very easy and low cost to maintain. Aircraft parts are readily available. The aircraft structure is easy to disassemble and assemble. The entire wing can be disassembled and replaced.

Douglas produced successful commercial and military aircraft before and after the war. By 1945, the company had produced nearly 30,000 aircraft of all models and the number of employees had grown to 160,000. The company produced many types of aircraft, including the DC3, DC-6 (1946), and DC-7 (1953). The DC-8 was produced in 1958, with a total production of 556 (1958-1972). From 1965 to 1982, Douglas produced 976 DC9s. The DC-9-80 was later redesignated as the MD-80. In 1987 Shanghai Aviation Industrial Corporation (SAIC) was licensed by McDonnell Douglas to assemble MD82. By October 1991, all 25 MD82s had been assembled.

The famous Boeing B737-100 made its first maiden flight in 1967 and entered commercial service in 1968 to compete with DC9. According to a report, as of January 2025, Boeing has delivered 11,965 B737. It is worth noting that very few airlines were interested in buying the first-model B737. A similar situation also happened to the first model of Airbus A300. Airbus was unable to find any buyer for many of the new aircraft that had already been produced (White tail airplane). Airbus was near collapse before it received financial help from its partner governments.

In the 20th century, the European aircraft manufacturers had technology more advanced than the US manufacturers. As the European commercial aircraft market is too small for too many manufacturers to sustain, many of them had to cease operation even though they had some sales in the overseas market. The US airlines prefer to buy locally made aircraft for better logistics and technical support.

21.2 USA Federal Aviation Administration (FAA)

The FAA's main roles in the US include regulating civil aviation to promote safety, encouraging and developing civil aviation, including new aviation technologies, and developing and implementing aircraft noise abatement and other civil aviation environmental impact control programs. In terms of safety regulation, the FAA issues and enforces regulations and minimum standards for aircraft manufacturing, operation, and maintenance. Within the scope of research, engineering and development, the FAA develops systems and procedures necessary for safe and effective air navigation and air traffic control systems. The FAA helps develop better aircraft, engines, and equipment, and tests or evaluates aviation systems, equipment, materials, and procedures. The FAA works with foreign civil aviation authorities.

If the FAA believes there is a problem involved in the safe operation of an aircraft, the FAA will issue an Airworthiness Directive (AD) to resolve the problem. For example, both Lion Air Flight 610 on October 29, 2018, and Ethiopian Airlines Flight 302 on March 10, 2019, operating B737 Max8 during takeoff went out of control and crashed. The two accidents killed 346 people. The FAA issued an injunction to stop the B737 Max 8 aircraft from continuing operations until the cause of the crash is determined and resolved. Each new aircraft model must be certified and approved by the FAA before it is allowed to fly. China has the Civil Aviation Administration of China (CAAC) and Europe has the European Union Aviation Safety Agency (EASA) to perform similar duties. The newly built C919 model built by COMAC must be certified and approved by the CAAC before it can begin commercial operation on May 28, 2023.

21.3 EUROPEAN AIRCRAFT MANUFACTURERS COULD NOT COMPETE WITH THE US

With the introduction of jetliners in the 1950s, several European aircraft manufacturers decided to build jetliners for the medium and long haul market to compete with the US. However, many European aircraft manufacturers failed to compete as follows:

- Aircraft manufacturing in the Netherlands

In the early 1960s, the Dutch company Fokker started launching its F28 twin-jet short-haul jet as an upgrade to its popular F27 Friendship twin-turboprop. Fokker received support from the Dutch and West German governments, partnered with the British Short Brothers, and acquired the West German MBB and VFW. Each partner was responsible for a different assembly, with the final assembly in the Netherlands. The aircraft was powered by two Rolls-Royce turbofans. By 1987, Fokker had built 241 aircraft. In addition to commercial users, the F28 was also used as a transport vehicle by the militaries of several countries. Finally, the Fokker F-28 never really became a major competitor to the DC-9 and B737. Founded in 1919, Fokker declared bankruptcy on May 15, 1997.

- Aircraft manufacturing in Germany

The German VFW 614 aircraft was originally proposed in the early 1960s as a 36-40 seat aircraft concept. It was a twin-engine jetliner designed and built by the merger of the West German airline VFW and the Dutch aircraft company Fokker. It was the first jet-powered passenger aircraft developed and produced in West Germany. The anticipated sales and production of the aircraft in the early to mid-1970s did not materialize, and the program was canceled in 1977.

- Aircraft manufacturing in France

In 1967, the French company Dassault decided to compete with the B737, developing a similar aircraft with 140 seats. This is Dassault's first foray into the commercial jetliner market. The new aircraft was named "Mercure" and was powered by a pair of American Pratt & Whitney JT8D-15 turbofan engines. Mercure entered service with Air France International on 4 June 1974. Other airlines showed little interest in Mercure. As a result, only one customer ordered 10 aircraft. Eventually, the aircraft program was terminated.

The Caravelle is a French passenger jet produced by Sud Aviation. It was developed in the early 1950s and first flew on May 27, 1955. On 26 April 1959, the aircraft entered SAS commercial service in the Nordic country. A total of 282 were built. It was ordered by airlines on every continent and operated until 2005. The short-range aircraft was powered by two rear-mounted Rolls-Royce Avon turbojets. Sud Aviation has since stopped making commercial aircraft. SUD later became a partner of Airbus.

Concorde is a joint venture between Aerospatiale and BAC to build the world's first supersonic passenger jet. Construction of six prototypes began in February 1965, and the first test flight took off from Toulouse, France, on 2 March 1969. Concorde can fly at speeds of 1,347 miles per hour at an altitude of 60,000 feet while maintaining speeds of up to Mach 2.04. Delays and cost overruns increased the cost of the programme to between £1.5 billion and £2.1 billion before Concorde entered service on 21 January 1976.

On July 25, 2000, Air France flight from Paris to New York, crashed shortly after takeoff, killing all 109 people on board and four on the ground. It was the only fatal accident in Concorde's 27-year history. The plane ran over some debris on the runway, causing a tire to explode and disintegrate. Tire debris caused a fuel tank to leak and catch fire. Concorde was retired in 2003 after 27 years of commercial operation. Air France and British Airways remained the only customers with a total of 20

21.4 AIRCRAFT MANUFACTURING IN THE UNITED KINGDOM

The BAC1-11 prototype aircraft made its first flight on August 20, 1963, but it crashed on October 22, 1963. The first production BAC 1-11 was delivered to BUA Airlines on 22 January 1965. The aircraft had the most advanced avionics equipment. BAC produced a total of 244 aircraft. The model serves dozens of airlines around the world. In 1977, BAC merged with Hawker Siddeley to form British Aerospace Corporation.

Comet 4 was the world's first commercial passenger jetliner. Developed and manufactured in the UK by de Havilland. It has a four-person cockpit, staffed by two pilots, a flight engineer and a navigator. It first entered service with BOAC in 1952. However, within a year of entering commercial service, problems began to arise, with three Comets falling apart in midair within a 12-month period. Two of these accidents were caused by metal fatigue failures. Flaws in the design and construction were discovered, including improper riveting and dangerous stress concentrations around some square windows. A total of 114 Comets were built. The plane stopped flying in 1981.

The British-built VC10 is a long-range four-engine jetliner introduced in 1964, designed and built by Vickers-Armstrongs. It was built to compete with B707 on transatlantic flights and served BOAC, the Royal Air Force and several African and Middle Eastern airlines. After the last aircraft was delivered in February 1970, the production ceased and a total of 54 VC10s were built. According to the general guidelines required for the production of a new aircraft, orders of more than 200 aircraft are necessary to break even. As a result, large aircraft manufacturing like the VC10 in the UK could not meet the necessary conditions to break even in a small market. From the history of commercial aircraft manufacturing, although Britain and France aircraft manufacturers are more technologically advanced than their competitor of the US, British and French aircraft do not have a foothold in the large American market. Therefore, the Comac C919 aircraft business needs China's large market to have a chance of success.

21.5 AIRCRAFT MANUFACTURING IN JAPAN

On April 1, 2008, Mitsubishi Aircraft Corporation began to develop the MRJ (later renamed Spacejet) as a 70–90 passenger regional jet with certification for 2012. Flight tests were scheduled for late 2011. The project required to break even at 300-400 aircraft production to pay for the investment. The project was delayed due to design changes. The first delivery was delayed until 2020. Due to the epidemic and various other technical modification and debugging problems, as well as internal management conflict, the project faced further delay. Due to financial problems, the company laid off 95 percent of its workforce in April 2021. On February 6, 2023, Mitsubishi Heavy Industries terminated the Spacejet project after 15 years and 1 trillion yen expenses. The project was a failure due to the small domestic aviation market and lack of foreign customers.

21.6 Many American Aircraft Manufacturers Withdrew From the Market

After World War II, there were several commercial aircraft manufacturers in the US, and competition in the market was fierce. Unless the aircraft manufacturer receives a large order for the first model of the new aircraft produced, the manufacturer will not survive. To design a modern commercial aircraft, manufacturers need to hire 5-6 thousand aircraft design engineers and deal with thousands of domestic and foreign OEM equipment and parts suppliers. The aircraft must have the right engine to fit the airline's aircraft operation requirements. The airlines' requirements relate to the range, capacity, fuel consumption, speed, and price of the aircraft.

After the first model aircraft is produced, the manufacturer needs to retain the engineering designers to continue upgrading the existing model and design the next model to meet the future requirements of the airline customer. Due to the intense competition among all manufacturers, aircraft manufacturers will lose money unless manufacturers can secure enough initial customer orders to break even. Here are the failures of three US commercial aircraft manufacturers:

- General Dynamics: Convair 880 was the first aircraft built by General Dynamics. It was an American narrow-body jet designed to compete with Boeing 707 and Douglas DC-8. During production from 1959 to 1962, only 65 Convair 880 were built. General Dynamics eventually withdrew from the airliner market after the 880 project was deemed a failure, costing the company about $185 million.

- Lockheed: Lockheed was founded in 1926. In the 1960s, Lockheed began developing the L-1011 wide-body aircraft. The L-1011 was to compete in the same market as the McDonnell Douglas DC-10. Delays in the development of Rolls-Royce engines caused the L-1011 to fall behind the DC-10. The Rolls-Royce RB211 engine also had a number of problems, which eventually led to Rolls-Royce filing for bankruptcy in February 1971. In 1971 Lockheed was heavily in debt, requesting a loan guarantee from the US government to avoid bankruptcy. On March 15, 1995, Lockheed and Martin Marietta were merged and Lockheed ceased production of commercial aircraft. Lockheed produced only 250 L1011.

- Doulas Co. : Defects in the design of early MD10 cargo doors resulted in a poor safety record in early operations. On May 25, 1979, American Airlines Flight 191 crashed immediately after taking off from Chicago's O 'Hare Airport. All 271 people on board were killed. FAA grounded all MD10 aircraft. On July 19, 1989, United Airlines Flight 232 crashed in Sioux City, Iowa with 111 people killed, and the aircraft was destroyed. As of September 2015, there had been 55 accidents involving MD10 with 1,261 deaths. In August 1997, McDonnell Douglas was acquired by Boeing. Since then, Boeing had monopolized the US commercial aircraft manufacturing industry.

21.7 CATIA AIRCRAFT
DESIGN SOFTWARE

In 1977, the French company Dassault created a computer aided three-dimensional Interactive Application (CATIA) software for aircraft design applications. The software drawings can be turned, allowing the designer to see the shape of the model front, back, left and right. CATIA can calculate whether various components can be fitted into the narrow storage space of the aircraft. It can save aircraft design time, and also improve the level of engineering precision. Aircraft design data can be stored forever and continued for future aircraft design processes and improvements and modification to existing aircraft.

CATIA can also be used by different design offices in different regions. For example, Airbus aircraft parts were designed by several European countries, and finally all aircraft parts were assembled in France. From the design drawings of various design offices, with CATIA and CAM (Computer Aided Machine Manufacturing) manufacturing process, every aircraft component and part can be manufactured automatically and precisely in any factory. Using CATIA, it is easy to modify any part of the aircraft. According to reports, China's Xi 'an Aircraft Manufacturing Group also uses CATIA to design and develop the JH-7A fighter bomber. CATIA is also used by other industries like ship building, architecture and building designs, and other engineering product designs.

21.8 History of Airbus

Airbus was founded on December 18, 1970. It is a corporation consisting of France, the United Kingdom, Spain, and Germany. As the member countries speak different languages and have different cultures, union rules and legal systems, Airbus management faced many challenges and operational problems at the beginning. Eventually, Airbus decided to use English as the Airbus international working language. Toulouse, France, is the headquarters of Airbus. Installed on the first Airbus aircraft model A300 were many parts, accessories, components, and engines imported from the US. The aircraft electronics operating system was designed in collaboration with McDonnell Douglas.

There had been a number of internal problems at Airbus that affected sales of the A300. Members of the Airbus alliance lack investment funds. Although, the first aircraft A300 model was fitted with an advanced aircraft operating system, fiberglass material to reduce the aircraft's weight, a two-person cockpit concept, and the first wide-body aircraft design, there was lukewarm reception by airlines around the world. Many airlines had no faith in a brand-new aircraft model from a brand-new multinational aircraft manufacturer in Europe. Many new aircraft sat on the tarmac awaiting buyers. The Airbus A300 began commercial service on May 23, 1974 with Air France.

Despite Britain's partnership with Airbus, no British airline ordered a single A300 aircraft. For years, British airlines preferred to buy American-made aircraft. When Airbus later hired an American as head of aircraft sales, in December 1977, he found his first US customer, Eastern Airlines (EA). However, the airline did not have sufficient funds to pay for the A300, and Airbus allowed EA use-first-pay-later deal. It was an ingenious marketing strategy. Later, EA ordered 23 airplanes. Pan Am (PA) followed the EA order shortly afterward with a deal for 13. American Airlines (AA) ordered 25. By 1979, Airbus had received a total of 256 orders.

Airbus's initial operation was marked by years of bumpy experience and challenges. After-sales product support services were not ideal, and there was a lack of suitable engineering representatives to help the airlines. There were lots of issues concerning on-time spare parts support. These teething problems were expected by a newly established aircraft manufacturer, especially a multinational company with different nationalities and having to deal with many suppliers from different countries far from the home base.

Building on the success of the A300, Airbus launched the A320 in March 1984. It was the first commercial aircraft to use a digital flight-by-wire control system. Airbus cockpit's aircraft flight management system with advanced electronic systems allows pilots to fly different families of Airbus aircraft with only flight simulator training, saving airlines the cost of re-certifying pilots for different models of Airbus aircraft. Compared with Boeing, Airbus A320 design is more innovative than Boeing, which is more conservative.

Airbus has taken advantage of the cockpit commonality of Airbus different models to reduce training costs for pilots. Airbus designed its aircraft to use fly-by-wire flight control systems to make it safer to operate. The new system eliminates the use of a hydraulic system controlling the movement of the aircraft, thereby reducing the weight of the aircraft. Older aircraft design utilizes mechanical and hydraulic mechanical flight control systems that are relatively heavy and require the flight control cables to be carefully connected to the aircraft via a system of pulleys, cranks, tension cables and hydraulic tubes.

The A320 was a huge commercial success. Airbus was lucky to select the CFM56 engine to power the A320 aircraft. The engine was first produced in 1974 and is a proven mature engine. As of April 2025, according to the report, a total of 12014 Airbus A320 family aircraft have been delivered with 19,234 orders placed in total. In comparison, Boeing has delivered 15156 Boeing 737 aircraft since its commercial operation in 1968. After 50 years of competition, Airbus has finally outsold the giant Boeing in global sales. Boeing's B737 reputation is now being questioned by airlines because of aircraft design and build quality problems that have caused accidents. Many airlines have cancelled B737 orders and switched to buy Airbus aircraft.

Airbus has opened an assembly plant in Tianjin, China, to build the A320 family of passenger planes, and is also building a $350 million parts manufacturing facility in Harbin. The facility will produce composite parts for A350, A320, and future Airbus models. Chinese partners such as Harbin Aircraft Industry Corporation, Hafei Aviation Industry Co LTD and Aviation Industry Technology Corporation of China hold 80 percent of the facility. Airbus holds 20 percent.

Airbus airplanes have been proven safe and reliable, never had a customer complain about any quality problem on its newly built airplane, unlike debris that had been found inside Boeing newly built airplanes. The quality problem is Boeing's Achilles heel. Comac is to appreciate that Airbus partners did not start from scratch, but had years of aircraft manufacturing experience. The Chinese government needs to pull out all the stops to aid Comac and other domestic OEM suppliers of aircraft components and parts. This process may require

15–20 years of financial assistance until Comac can build aircraft without the need for foreign supplied equipment and parts.

21.9 THE SUCCESS OF CFM56 ENGINE

CFM is a 50-50 joint venture between Safran Aero Engines (formerly Snecma) of France and General Electric Aviation (GE) of the US. The cooperation between the US and France as business partners is a miracle, as Safran is a partly state-owned company. The first production CFM56 engine ran in 1974. The joint venture had not received a single order in the first five years and was only two weeks away from being dissolved when a stroke of luck happened. The CFM56 project survived when Delta Air Lines and United Airlines selected the CFM56 to reengine their popular DC-8-60 in 1979 for noise reduction and better fuel consumption. Soon after, it was selected for re-engining on the US Air Force Boeing KC-135 tanker.

The opportunity was given to CFM by Pratt & Whitney (PWA), which refused to design a new engine to replace the noisy JT3D engine produced in 1958. Because the JT3D engine was selling very well in the market, PWA did not see the need to design a brand-new engine to replace the JT3D engine. The new CFM56 is 23 percent more fuel efficient than the JT3D engine, thus reducing operating costs and extending the DC-8-60 aircraft range.

In the 1970s, noise pollution was a hot topic and CFM 56 was a quieter engine. In the end, McDonnell Douglas chose the new CFM 56 to power a new stretched version of the DC8-70. The CFM 56 grew rapidly in sales and was selected by Boeing to power the new B737. Since then, Pratt & Whitney has lost the engine leadership position. To date, more than 33,400 CFM56 engines have been delivered to power the A320 and B737 families of aircraft, making it the most popular engine in commercial aviation.

21.10 Boeing Company

The Boeing Company was founded by William Boeing in 1916 in Seattle, Washington. Boeing is one of the world's largest aerospace manufacturers. The first engineer at Boeing Aircraft Company was Wang Suke, a Chinese graduate of the Massachusetts Institute of Technology. He was hired by Boeing in May 1916 as chief engineer. He designed the Boeing Model 2, which was Boeing's first financial success. In 1916, the US Navy purchased 53 Model C trainers from Boeing and the Army purchased two side-by-side seating land aircraft, giving Boeing the first financial success. His contribution lied in his expertise in analyzing aircraft wind tunnel data. Wang was known as the "Father Of Boeing." Because he was banned by the US government for not participating in flight tests, he resigned and returned to China.

Boeing executives have significant political influence in American politics. Boeing is a publicly traded company whose board includes former military executives. Aviation history shows that American aircraft manufacturers have two advantages: a large market size and the availability of engines to fit any aircraft design. European aircraft manufacturers have only one engine maker, Rolls-Royce, with a small customer base. The viability of private engine manufacturers in the US depends on the huge support of the government to spend vast amounts of money on R&D and supply of engines to the vast fleets of air force jets.

Since 1958, Boeing has built the following commercial aircraft: aircraft model: first aircraft operating year, total production figure

B707, 1958, 865
B727, 1964, 1832

B737, 1968, 15156

B747, 1970, 1562

B767, 1982. 1263

B757, 1983, 1050

B777, 1995, 1748 (as of March 2025)

B787, 2011, 1174 (as of March 2025)

To avoid the cost of designing and building an entirely new aircraft model that would require years of testing and experimentation, Boeing resorted to stretching or shortening the length of the original aircraft model to produce a derivative aircraft that would not need to go through the entire same lengthy certification processes. Many of the original aircraft components can be used in the derivative aircraft, thus reducing the cost of the derivative aircraft. The only changes required were a new engine with higher or lower thrust, and the landing gear of the newly derived aircraft. In addition, new derivative aircraft do not require the pilot to apply for a new pilot's license, as long as he has a license to fly the original aircraft model or any other derivative aircraft. This will reduce pilot training costs for the airlines.

21.11 TWO BOEING B737-MAX 8 AIRCRAFT CRASHED

Boeing suffered from two fatal plane crashes involving the B737-MAX 8 aircraft. Lion Air Flight 610 and Ethiopian Airlines Flight 302 crashed on October 29, 2018, and March 10, 2019, respectively, killing 346 people in total. The US Federal Aviation Administration ordered all Boeing 737 MAX 8 aircraft to stop flying worldwide, immediately affecting 387 aircraft operations worldwide. Airlines suffered heavy financial losses. Boeing's 737 MAX 8 has been fitted with a new Maneuvering Characteristics Augmentation System (MCAS) designed to help pilots avoid mistakenly raising the plane at too high an angle of attack. However, in some flight situations, the MCAS can unexpectedly lower the nose rapidly and the pilot cannot control the airplane, causing the airplane to dive or crash.

There is a fundamental flaw in the initial design of the MCAS system, which relies on taking only one angle of attack sensor (AoA) signal to trigger the MCAS action. Boeing later updates the MCAS software to now compare data from both AoA sensors, enhancing redundancy through software. If the sensors disagree by more than 5.5 degrees, MCAS will not activate. Additionally, MCAS is modified to activate only once per high-angle-of-attack event instead of repeatedly as in the past, and pilots are given more direct control to override it as required.

Some airlines believe that the B737 MAX 8 design shortfall was due to the haste of Boeing wanting to compete with Airbus 320neo quickly without enough time to complete proofing the design and test flights. Proofing mitigates risks of post-launch errors, saving time and resources while ensuring the final output aligns with expectations. Obviously, the original MCAS had not been subject to a complete proofing exercise. It is inconceivable that the pilots cannot override the erroneous MCAS action. In December 2020, the 737 MAX 8 with FAA approval resumed commercial operations. However, in April 2021, an electrical problem occurred that affected the safe flight of more than 100 aircraft. Lack of safe electrical grounding could lead to the failure of various electrical systems, such as the engine anti-icing system and the auxiliary power unit. It was the latest problem to plague Boeing's top-selling aircraft.

The two B737 MAX 8 crashes serve as a valuable reference to Comac that the C919 must be designed and tested according to international safety standards and not to neglect complete aircraft design proofing and test flight programs and not to rush to sell the aircraft worldwide. It would be advisable to fly the C919 only within China for the first 3–4 years and wait until the aircraft teething problems are ironed out before selling it abroad. New model aircraft always encounter unexpected problems that only surface during commercial operation. There are issues of technical and spare parts support from foreign suppliers, and practical maintenance problems requiring new tools or equipment. The C919 will be closely watched by Boeing, Airbus and global airlines. The successful operation of the C919 aircraft will positively help future C929 model sales. There is no lack of customers in the aviation world when Comac C919 is proven to be a reliable aircraft.

Boeing has had an ongoing systemic quality control problem. For decades, the FAA has commissioned employees of Boeing and other large aerospace manufacturers to sign off on certain aspects of aircraft design on the agency's behalf. Now regulators are considering bringing in a third party to independently oversee Boeing's inspections and quality control, underscoring Boeing's loss of confidence. In the commercial aviation industry, getting the quality details right is literally a matter of life and death. A former Boeing executive said the financial cost of grounding the 737 Max 9 could far outweigh the long-term loss of confidence it would cause.

Airbus has focused on maintaining the highest quality of aircraft manufacturing, avoiding a string of defects like Boeing, while maintaining higher production volumes than its rivals. A very important point for all business dealings, whether it is a product or a service, is that everyone is selling trust to the customer. Boeing's series of quality failures have seriously eroded customers' trust in Boeing. Comac can learn a lesson from Boeing's quality problems. The same fate befell McDonnell Douglas with its MD10 suffering from 55 accidents involving 1261 fatalities because of quality problems.

21.12 Difficult Birth of Boeing Dreamliner B787

In 2006, Boeing called the B787 Dreamliner a "game-changer" because the company was using a significantly different approach to aircraft design that it said would transform aviation. Analysts estimated that Boeing and its partners would invest $8 billion to develop the B787, the company's first new model commercial plane in a decade. It was initially expected to go into service in 2008. Due to several delays, B787 type certification was approved only in August 2011, and the first B787 entered commercial service on October 26, 2011, with ANA. Then more problems befell the Dreamliner.

Problems affecting birth of B787 as followed:

- In October 2007, Boeing said that it would delay initial deliveries of the 787 Dreamliner by six months, to late November or early December 2008. The delay was caused by the problems with Boeing's global chain of suppliers as well as unanticipated difficulties in its flight-control software

- By September 2009, the effects of Boeing's outsourcing problems had taken a huge toll. The company's chief, W. James McNerney Jr., conceded that Boeing lost control of the manufacturing process by farming out more design and production works than ever before and not keeping close tabs on the suppliers.

- Boeing said it would delay deliveries of the B787 until the first quarter of 2010, nearly two years behind the original schedule. The company attributed the delay to a strike by the International

Association of Machinists and Aerospace Workers, and also said it had encountered problems with fasteners.

- Boeing said the first Dreamliner would not be delivered until the middle of the first quarter of 2011. The company cited the unavailability of Rolls-Royce engines

- Boeing halted test flights of B787 after an onboard fire forced an emergency landing in November 2011. The smoke entered the cabin from an electronics compartment in the rear of the plane. Boeing said it would push back the first delivery of the B787 to the third quarter of 2011 to redesign some affected parts.

- B787 type certification was approved by FAA in August 2011

- In early December 2012, a United Airlines B787 flight experienced one electric generator failure. On the same day, the FAA ordered inspections of fuel line connectors on all B787s, warning of a risk of fuel leaks and fires.

- On January 16, 2013, the FAA issued an emergency airworthiness directive requiring all US airlines to ground their Boeing 787s until electrical system modifications were made to reduce the risk of batteries overheating or catching fire.

- In 2013, the FAA said it would review the safety of Boeing's 787 aircraft. The entire Boeing 787 fleet has been grounded after the FAA ordered a review of the plane's design and manufacturing.

21.13 Boeing Subject to Scrutiny for Poor Manufacturing Quality

Boeing has been plagued by quality problems for more than 40 years. Many new aircraft were delivered from factories, with many repairs performed on the new aircraft due to mistakes made by production workers. Compared to Airbus, every new aircraft is flawless without repair. In addition, no workers leave any foreign objects on newly delivered Airbus aircraft, whereas such careless mishap happened in the Boeing factory. In 2009, Boeing built a new facility near the city of North Charleston (population about 100,000) that was hailed as Boeing's most advanced manufacturing center and one of the most advanced in the world. But in the decade since, the factory that makes the B787 has been plagued by poor production quality and poor oversight, threatening aircraft safety. A New York Times review of hundreds of pages of internal Boeing emails, company documents and federal records, as well as interviews with more than a dozen current and former employees, uncovered a culture that often placed a premium on speed of production over quality. Faced with long production delays, Boeing forced its employees to quickly produce the B787, sometimes ignoring production problems raised by workers. Qatar Airways said it won't accept airplanes from this factory.

Boeing is the only commercial aircraft manufacturer in the US, and so management becomes complacent and apathetic because it monopolizes the commercial aircraft market in the US. As a result, management becomes lax and less involved in the management and improvement of product quality. Rich Mester, a former Boeing technician, did an inspection of the new airplane before it was delivered to the customer. He had found floating objects inside the airplane, such as sealing hoses, nuts, things used in the manufacturing process. Some staff members even found a ladder and a string of lights near the horizontal stabilizer gear in the tail of an airplane. The cabin of a B787 built for American Airlines suffered severe flooding, and the seats, ceilings, carpeting, and electronics had to be replaced. In an effort to meet deadlines, managers sometimes downplayed or ignored quality issues, as reported by the current and former employees. Top managers urge internal quality inspectors to stop documenting defects.

On November 18, 2020, the Federal Aviation Administration approved the return to service of the B737 MAX 8 , whose grounding cost Boeing an estimated $20 billion in fines, restitution and legal fees, as well as consequential damages, including the cancellation of 1,200 orders worth more than $60 billion. In December 2020, the B737 MAX8 resumed commercial operations and deliveries, but in April 2021, a new electrical issue affected nearly 100 planes, prompting another partial ban and further scrutiny of production issues on Boeing airplanes. It seems Boeing quality problems have become a terminal disease.

21.14 CHINA'S CIVIL AVIATION INDUSTRY HISTORY

The People's Republic of China was founded in 1949. Due to the friendly diplomatic relations between the two countries, China followed the Soviet Union's aviation system and regulations. At that time, European and American countries banned the export of aircraft to China, and hence China can only use the Soviet-made passenger aircraft to fly internally. On February 21, 1972, President Nixon of the US arrived in China for an official visit. In the same year, China ordered 10 Boeing B707 aircraft. In 1973, China received its first Boeing B707 and began international flights. On January 1, 1979, China and the USA established diplomatic relations. In 1978, Deng Xiaoping introduced an open market economy policy. In 1987 Shanghai Aviation Industrial Corporation (SAIC) was licensed by McDonnell Douglas to assemble MD82. By October 1991, all 25 MD82s had been assembled. In March 1992, China and McDonnell Douglas started the joint-production of MD90, and the degree of localization of the airframe reached 70%. However, in August 1997, after McDonnell Douglas was acquired by Boeing, the cooperation between the two sides came to an abrupt end.

In 1988, China planned to cooperate with the German MBB company to produce a 75-seat regional aircraft. Over the next few years, 176 Chinese engineers were trained in Germany. The collaboration resulted in technology transfer worth DM 26 million, but the project itself did not make substantial progress. Later, the Aviation Industry Corporation of China signed an agreement with Airbus and Singapore Technology Company to cooperate in the development of a 100-seat regional aircraft AE100 project. On grounds of disagreement, Airbus announced the end of the AE100 project.

In 1960, the Soviet government unilaterally decided to recall all of their 1,390 Soviet experts working in China, and stopped 900 Soviet experts who had already applied to work in China. At the same time, the Soviet side tore up 12 agreements signed by the two governments and one agreement signed by the scientific academies of the two countries, as well as cancelling 443 expert contracts and 257 scientific and technological cooperation projects. Since then, the door for China to continue to obtain aviation technology from the Soviet Union was closed.

As early as the late 1950s, China's Xi 'an Aircraft Industry Corporation began to reference the former Soviet Union's medium jet bomber TU-16 development to build a similar bomber. After being stalled for various reasons, the aircraft production program was restarted in 1963, and the first H-6 prototype was completed in October 1966 for static testing. On December 24, 1968, the H-6, equipped with China's domestic turbojet 8 engines, made its first flight successfully. China has thus started its own domestic aircraft to fulfill its dream. Finally, China successfully produces the C909 by Comac with a maiden commercial flight on June 28, 2016, followed by C919 commercial flight on 28 March 2023 by China Eastern Airlines.

21.15 China's Commercial Aircraft Manufacturing Industry

To design and manufacture the most advanced large commercial aircraft, China needs to have sufficient capital to invest in equipment, instruments, laboratories and 5,000-6,000 experienced engineers, technicians, and pilots to have a chance of success. The first prototype will require at least four years of test flights to ensure the aircraft is functional and meets airline and international aviation flight requirements and standards. After the aircraft is sold, it will have to invest further in continuous improvement of the product. The aircraft needs to operate 30–40 years safely.

Product support service is one of the most important commercial aviation requirements that involves many suppliers and different countries. Chinese private companies simply do not have the capital, technology, and capacity to take on this extremely risky investment. As a result, Comac was chosen by the government to design and build China's first mid-sized commercial aircraft. China is fast becoming the world's largest economy, so it cannot continue to rely on foreign commercial aircraft to meet the country's needs. Comac can only compete successfully with rival aircraft manufacturers in terms of favorable price, reliability, comfort and safety, and with lower ownership and maintenance costs.

Comac needs to understand that airlines not only sell seats to passengers, but also provide a complete package of passenger services before, during and after the flight. The entire senior airline management is involved. Although an aircraft is just a means of transportation, passengers also need a safe flight, a comfortable cabin environment, inflight entertainment, on-time departure and affordable ticket price. Airline operations are a series of activities carried out in the supply chain system that cannot be interrupted anytime, anywhere. Only by continuously developing better passenger aircraft that meet the actual needs of the market can an aircraft manufacturer remain profitable. Every new aircraft development project requires about five to six thousand engineers to work through the completion of the new project. After the new project is completed, the entire design team must continue to work on the next new project for a derivative aircraft, which may be the first version of a new shortened or lengthened model aircraft.

Comac project engineering departments must explore early on the needs of airlines for future operations and aircraft upgrades. In addition, Comac must work with engine manufacturers on their development plan and cost of new engine models, continuous upgrade of engine reliability and reduction of fuel consumption. At present, C909 and C919 rely on many American component and avionic system suppliers. Such aircraft manufacturing cannot be maintained for long because of international geopolitics. China must change its immigration policy to attract experienced European scientists and engineers from the UK, Germany, Russia, and Ukraine to work and live in China with permanent residency to help accelerate the development of commercial aircraft in China. The US did this after World War II, and China can do the same.

It took Airbus nearly 50 years to beat the US in the international commercial aviation market, and Airbus engineers have years of prior experience designing and building commercial aircraft. Airbus did not start from scratch. Comac designers lack practical experience in the operation and maintenance of commercial aircraft. In the long run, China must produce its own commercial aircraft independently , at least for its own huge domestic market. It doesn't matter if it takes China 50 years to catch up with the US in international markets. Do not follow the example of Boeing's rush to design the B737 MAX 8 to compete with Airbus while ignoring the safety design rule. Safety is paramount in aviation business. The McDonnell Douglas MD10 failed due to multiple crashes.

The aircraft manufacturing business is a business integrator that assembles hundreds of thousands of parts from different suppliers from different countries. Its business is like Huawei's smartphone assembly, involving many suppliers of components such as touch screens, motherboards, buttons, batteries, cameras, various types of screws, motor vibrators, speakers, memory chips, screens, housing, antenna, mic, controllers, gravity sensors, processors, chips of various apps. At present, China is lacking experience in designing and manufacturing many aircraft mechanical and avionic components and software systems. It will be at least another 20–30 years before China can produce all the components it needs to expand in future to compete with European and American aircraft manufacturers. There are still a few American parts on Airbus airplanes. Some of these experienced American suppliers have been in the aviation business for decades.

The US engine manufacturers have strong support from the US government's Air Force and the domestic aviation market, as well as decades of valuable experience in operations. US private engine manufacturer research and development has been quite high, supported by US government funding through military contracts. It is reported that the US Air Force has about 6,000 aircraft. At the same time, the US also sells thousands of military aircraft to foreign countries. The US government buys aircraft on a cost-plus basis. So, profits are guaranteed for the US aerospace industry. Military aircraft business is a low-hanging fruit, more lucrative than commercial aircraft business. Senior executives in the military industrial complex and the senior politicians in the Congress have a close relationship.

21.16 THE FUTURE OF CHINA'S COMMERCIAL AIRCRAFT INDUSTRY

C919 is the first medium-sized passenger aircraft of Comac launched in 2008. It is China's first civil jet designed in accordance with the international aviation standards and developed by itself and has independent intellectual property rights. It has 158–168 seats and a range of 4,075-5,555 kilometers. The C919 is equipped with twin-engine turbofan CFM LEAP-1C engines from CFM International, a joint venture between Safran Aircraft Engines and GE Aviation. The aircraft has an economic life of 90,000 flight hours. C919 reportedly has a total of 1 million parts and accessories and 80 kilometers of electrical wiring. Some of the 1 million parts and accessories are imported from abroad. More than 200 Chinese enterprises participated in the C919 development. Comac has two certificates: TC (Type Certificate) and PC (production license), which are Comac property rights.

If China is to establish itself as an independent aircraft manufacturer that does not depend on component imports from the US or the European Union, the government must provide financial aids to encourage thousands of private Chinese companies to invest in various sectors of the aviation industry. It may take 15–20 years of experience to reach a mature aviation industry. After World War II, the aviation industry of the US relied on many European scientists, engineers, and skilled immigrants to achieve today's achievements.

To be successful , Comac engineers must be proficient in the design of aircraft structures, engines, and practical knowledge in chemistry, materials, electronic flight control and automation systems, hydraulic systems, pneumatic systems, integration of various mechanical and electronic components and systems, hardware and software. One US engine manufacturer took nearly 8 years to proof-test its new invention of engine electronic fuel control to replace its mechanical fuel control, which was unreliable and often caused engine malfunction.

When Boeing B747 aircraft started commercial operation in January 1970, it ran into a lot of technical problems with the engines and the aircraft's fly-by-wire flight control system. Neither the engine nor the control system had been tested in commercial operation before. The engine intake fairing was deformed and the turbine temperature was too high. Many airlines operating the B747s experienced lengthy technical delays, sometimes more than 24 hours. In addition, the airlines must stock more spare parts and engines for replacement. The whole technical problem took 4–5 years to solve. It is not unexpected that the C919 could also take 4–5 years to iron out all the teething problems during the initial commercial operation.

According to report, C919 incorporates components from the following foreign suppliers:

- Liebher-- air conditioning
- Nexcelle -- engine's nacelle, thrust reverser and exhaust systems
- Michelin -- radial tires
- Liebherr and AECC -- landing gear
- UTAS -- power, fire protection and lighting
- Rockwell Collins -- cabin systems and avionics, fire system, in-flight entertainment system, electric power generation, traffic alert and collision avoidance systems
- Honeywell -- flight control, APU auxiliary power unit, wheels and brakes
- Parker -- hydraulics, actuators and fuel systems.

- GE Aerospace -- LEAP-1C engine, flight recorder, flight control system
- Safran SA – water and waste systems, thrust reverse control system.
- Crane – power transformer, flow meter, position indicator

21.17 Comac C919 Flight Tests

In April 2017, the C919 underwent a critical static failure test at the Yanliang Aircraft Strength Research Institute and was a complete success. According to reports, on July 26, 2017, Comac's third C919 aircraft took off from Shanghai Pudong Airport and flew 1,500 kilometers in less than three hours to land at Xi 'an Yanliang Airport, successfully. The airplane then conducted flight tests here to check for flutter, airspeed calibration, stability, and performance. The airplane must pass through all flight modes for performance at different air speeds, ambient temperatures, altitudes, wind conditions, flight loads and emergency situations.

The flight test also included turning off and relighting the engine while in the air. According to Comac's plan, it will take at least three years from the completion of the first flight test to the official launch of commercial routes. The flight test would be followed by other test verification and approval to final model certification. The three-year period also included flight tests to find any faults that needed to be rectified. By 2022, all six test aircraft had been put into certification flight test processes.

C919 is estimated to cost between US$90 million and US$100 million, compared with Airbus's A320neo at $111 million and Boeing's 737 MAX 8 at $121 million. On Nov 27, 2020, the C919 received the type inspection authorization from the Civil Aviation Administration of China, which meant that the design of the aircraft had been completed and verified, and there could be no big changes in the structure. The CAAC issued the C919 airworthiness certificate to Comac on Sept. 29, 2022. The first C919 aircraft was delivered to China Eastern Airlines in Shanghai on Dec. 9, 2022. The first four aircraft on the order would be delivered in 2023. It is reported that Comac has received orders for at least 1,000 C919s, including 20 from GECAS of the United States, 7 from Pren of Germany and 7 from Metropolitan Air of Thailand.

21.18 COMAC PRODUCT SUPPORT SERVICE

As with all new aircraft models produced in the US and Europe, the C919 is bound to have early teething technical problems. Problems can be related to premature failure of aircraft components or systems, maintenance errors, pilot error, short supply of spare parts and long lead times of repair, inexperience of product support engineers, communication problems, and errors in technical maintenance manuals. Comac should hire experienced aviation English speaking technical writers to proofread the flight operation manuals and aircraft maintenance manuals. Direct Chinese-English translation is not recommended as wrong choice of word in the written expression that might be misunderstood or misinterpreted. These problems may affect the safety issues of the aircraft operation.

Comac should learn a lesson from Airbus's experience after its first passenger A300 aircraft that was put into commercial operation. In the first few years, there were many problems during the international commercial operations: a) problems with on-time supply of spare parts, b) after-sales service could not meet the international standard level, c) many internal product support engineers having problems in communicating in proficient English level, d) problems in the English expression in aircraft operating manual. All these teething problems created unhappiness in the airlines, suffering from many airplane delays. It took several years for the Airbus service learning curve to improve. Airbus had internal problems because of different cultures of partner nationalities. Airbus was on the fringe of being financially dissolved until the respective partner governments provided financial aid in time.

It is prudent for Comac to concentrate on the C919, spending the first 3–4 years operating only within China so that it is easier for Comac to provide technical and spare parts support promptly and to rectify any mistakes in the airplane operating and flight manuals. Comac will have a hard time to deal with technical problems experienced by a foreign airline during the initial infant period of C919 operation. Teething problems with new flight management system software can happen, and software upgrade can take a long time when the related test software in the laboratory is outsourced to another company. Comac needs to build a comprehensive inventory of spare parts initially to support every airline operation around the world. Moreover, Comac and its foreign suppliers must have close commercial and technical relations, and must collaborate and work together to help airlines solve critical problems quickly.

Comac product support department must daily keep track of every airplane operating problems experienced by the airlines. Any repeated technical problem must be investigated and fixed to be provided by issuing a service bulletin. Comac should also monitor the unscheduled removal rate of every component to ensure that it remains at an acceptable reliability level. An annual conference should be held at Comac headquarters to discuss aircraft operational issues and solutions, as well as the cost of aircraft and engine maintenance per flight hour. Aircraft operating statistics are to be published monthly, including aircraft scheduling reliability, flight delays and cancellations, aircraft grounding , engine in-flight shutdown and other operational technical issues.

21.19 Aircraft Turbofan Engine Design and Manufacture

Turbofan engines that can safely and efficiently power large commercial aircraft carrying hundreds of passengers each time is difficult because it involves so many aspects of science and technology, manufacturing capability, durability, safety, commercial viability, and reliability for thousands of flights. Developing a commercially viable turbofan engine requires investing a lot of financial and human resources. The vast capital investment must be continuously utilized to produce marketable engines in a competitive environment. The manufacturer must have the capability to produce engines of various thrust ratings for applications of different models of aircraft configurations. Moreover, the engine manufacturer also depends on many subcontractors to supply special components and parts such as sensors , switches, bearings, pumps, wires, valves etc. to assemble an engine

US engine manufacturers have an advantage of a large domestic civilian market and a lucrative military engine market, aided by substantial government financial support for R&D. Rolls and Royce (RR) in the UK faces stiff competition from the US engine manufacturers, as it has a small domestic market in the UK and Europe. RR was fortunate to be selected as the sole engine supplier for Airbus airplanes A330Neo , A350 and Boeing 787. CFM was the only engine manufacturer for the first Airbus A320 airplane. Selecting a sole engine supplier will reduce the cost of airplane certification compared with two different engine models certification.

The Chinese government will endeavor to help develop a homegrown engine for the C919 in its national five-year plans. The government said in its 2021-2025 development plan that China will press ahead with the development of a turbofan jet engine CJ1000 for the Comac C919 narrow-body passenger airplane. Turbofan engines are difficult to build for the following reasons: engines operate at extremely high temperatures and pressures. The fuel burns inside the engine at a temperature up to 2000 °C. At present, China's technical expertise needs to build up in the following fields: precision turbine blade forging, powder metallurgy of turbine and compressor discs, and titanium alloy hollow parts molding, combustors that subject to most intense thermal and vibratory stresses, and to be made of high strength superalloys with refractory coating metals such as tungsten, molybdenum, niobium, tantalum, and turbine blades made of nickel-based superalloy incorporating internal air flow passages inside the turbine blades. China state-own company AECC had developed the CJ-1000A turbofan engine for C919 airplane.

In order for AECC to understand the performance of its new model CJ-1000A engine, it is necessary to trial install one CJ-1000A on one of the original test flight C919 aircraft to quickly establish more cycles to proof-test the reliability of the engine. The test engine will go through a total of 4000 cycles, cumulating from 10 times of 400 cycles each. The test engine should be removed at an interval of 400-cycle operation to be disassembled and checked for conditions in the shop. Any part found damaged should be replaced.

If the test engine can function 4000 cycles with reasonable wear and tear, it is considered to be successful. It is estimated that the C919 will fly 3000 hours per year and 4 cycles per day. 4000 cycles represents about 5–6 years of engine operation on the aircraft. This figure is very respectable for a reliable engine. When the CJ-1000A is mature, Comac can provide airlines with two different types of engines, one CFM LEAP-1C and the other CJ-1000A. The C919 and its future variants will have a long production line, as China's aviation market needs many C919 aircraft for its expansion of commercial aviation.

The C919 will be equipped with an engine condition monitoring system to monitor engine parameters such as compressor speed, vibration, oil pressure, oil temperature, exhaust gas temperature (EGT) and fuel flow to determine the deterioration of engine performance. The data will be sent to AECC headquarters for analysis of the engine performance. If a rapid deterioration in engine performance is detected, the engine must be removed for a thorough investigation to find the root cause of the deterioration. AECC needs to issue a service bulletin as soon as possible to solve this problem. Every engine will undergo periodic borescope inspections of the conditions of the turbine and compressor blades and vanes.

Chapter 22 Tourism in China
22.1 China's Tourism Industry

Tourism is an important industry contributing towards many countries' economies around the world. For China, tourism can enhance China's soft power, provide millions of jobs, increase foreign exchange earnings, help develop the country's tourism infrastructure and create cultural exchange between foreigners and local citizens. Many tourists travel to experience the culture, different traditions and cuisine of China. Tourism helps promote China's economic progress that leads to building more hotels, airports, roads and highways, development of parks, and improvement of tourist attractions.

It was reported that in the year before the pandemic, 5.539 billion domestic trips were made, and the total tourism revenue for the year was 5.97 trillion yuan. The overall contribution of tourism to GDP was 9.94 trillion yuan, accounting for 11.04% of the total GDP. In 2019 before the pandemic, the number of international tourists to China was about 145.3 million. Due to the Covid-19, the tourism industry was badly affected in 2022, with a total of 2.530 billion domestic tourist visits and 2.04 trillion yuan in domestic tourism revenue. In 2019, China's international tourism revenue was about US $131.3 billion, or US $94 per capita which was lower than many countries. Macau's economy is largely dependent on tourism services from its vast casino resorts. One third of Macau's GDP comes from tourism, and one third of its workers are employed in tourism. In 2024, China saw a significant rebound in inbound tourism with nearly 95 million inbound tourists in the first three quarters of 2024. For the same period, domestic tourists hit 4.23 billion. China introduced in 2024 visa free transit policy to many countries which can promote more tourists visiting China.

22.2 Scenic Spots in China

China has the following places of interest for domestic and international tourists:

- The Great Wall
- The Forbidden City
- The Summer Palace
- Temple of heaven
- Mogao Grottos, a 1,000-year-old man-made cave in Dunhuang, Gansu province, contains Buddhist heritage
- Kuzhen, located at the top of Mount Lushan, was once a summer resort for   European immigrants in southern China
- Heilongjiang Harbin International Ice Sculpture, with ice castle and magical snow scene
- Yabuli Ski Resort, Heilongjiang, the country's largest ski resort
- Longmen Grottoes, a parade of Buddhist statues and reliefs, near Luoyang
- The 2,200-year-old Terracotta Warriors guard the tomb of China's first emperor in the ancient capital Xi 'an
- Mount Tai, Shandong, is a sacred mountain with immaculate temples and pavilions
- Colonial buildings along the Huangpu River and Shanghai's skyscrapers on the Bund, Shanghai
- Xuankong Temple in Hengshan Mountain, Shanxi Province, a cliff temple, and a large number of Buddhist statues in the caves
- Leshan Giant Buddha, Sichuan, the world's largest carved Buddha at Emei
- Mountain in Sichuan province
- The Potala Palace in Lhasa, Tibet, built in 637

- Xishuangbanna Dai Autonomous Prefecture in Yunnan Province, home to China's most unique Dai ethnic minority
- Hangzhou, Zhejiang is famous for its West Lake, Zhangjiajie Forest Park, where the world's first elevator and Avatar movie locations

Michael Loong

22.3 Chinese Historical and Cultural Sites

- China's long history has left behind many cultural heritages. The 8,852 kilometers Great Wall is an iconic symbol of China and a typical historical site for tourists to see the magnificent structures that had withstood the elements for thousands of years. It was the greatest military defense construction project in the history of human civilization. Its history can be traced back to the Spring and Autumn Period and the Warring States Period more than 2000 years ago. More than a dozen sections of the Wall are now open to visitors, including bunkers and watchtowers at Badaling in Beijing, Laotong in Hebei province and Jiayuguan in Gansu Province.

- The Silk Road began in 130 BC when the Emperor of the Han Dynasty opened trade with the countries in Central Asia and Europe. Along the ancient Silk Road in Gansu Province, there are many caves filled with precious wall paintings and sculptures. The most famous is the Mogao Grottos, "a treasure house of Oriental art," with 492 caves and cliffs covered with murals and statues. There are 45,000 square meters of murals and more than 2,100 colorful statues. In the south, grottoes are represented by the Leshan Giant Buddha, carved on cliffs. With a height of 71 meters and a width of 28 meters, it is the largest sitting Buddha carved in stone, showing the carving skills of ancient craftsmen.

- Located in Henan province, the Shaolin Temple is the birthplace of Chinese Zen Buddhism and is famous for its Shaolin kungfu. It was founded in 495 AD. Five hundred Arhat murals from the Ming Dynasty and Shaolin kungfu paintings from the Qing Dynasty can be seen here.

- In Hubei, the Wudang Mountain, with 72 peaks and an area of 30 square kilometers, is a shrine to Taoism , preserving one of the most complete and largest ancient Taoist architecture in China.

- Located in western Sichuan, Mount Emei is full of ancient Buddhist temples and buildings and is one of the four sacred mountains of Buddhism in China.

- South of the Yangtze River, Suzhou and Hangzhou, known as "heaven on earth", with picturesque scenes of rivers, lakes, bridges, farm fields and villages.

- Well-preserved ancient cities like Pingyao in central Shanxi, are sites of the Yangshao and Longshan cultures of the Neolithic Age, dating back 5,000 to 6,000 years.

- Gulijiang in Yunnan is not only the center of Dongba culture of Naxi nationality, but also the meeting place of ethnic cultures such as Han, Tibetan and Bai. Built in the Song Dynasty, the city has many stone bridges, stone archways and dwellings, providing valuable information of architectural history and can be called "a living museum of ancient dwellings".

- In Xishuangbanna, Yunnan Province, the water-splashing Festival of the Dai ethnic group is a lively festival held in Spring time. People chase each other and splash water (a symbol of good luck and happiness), and there are activities such as dragon boat racing and peacock dancing.

- The Lugu Lake region, between Sichuan and Yunnan, is home to a matriarchal society of 30,000 local Motatus known for their "no-marriage" tradition, and has been called the last female kingdom on Earth.

22.4 TOURISM CONTRIBUTION TO NATIONAL ECONOMY

The Ministry of Tourism of China is to make a documentary film of all the scenic spots, natural beauty spots, historical and cultural sites and folk customs in China, which will be used as a global publicity material for China's tourism and help the development of China's tourism industry. The film can be shown in Chinese cultural centers abroad. It is reported that more than one billion tourists visit international tourist destinations every year. Tourism has become a major industry, accounting for 9.8 percent of global GDP . Tourism has become the world's third-largest export industry, behind fuels and chemicals, and ahead of food and automotive products. Macau earned an estimated US$32.6 billion in tourism revenue in 2023 for a population of 678,800 or US48,000 per capita, which is one of the highest in the world.

Before the Covid-19 in 2019, China had about 145.3 million international tourists with a revenue of about US$131.3 billion, or US $94 per capita, much lower than many other countries. China has not done enough advertising to attract more foreign tourists to visit the vast wonder and beauty of China's natural attractions, historical arts and crafts, and cultural diversity to earn a comparable per capita tourism revenue.

China should not only develop international tourism, but also develop domestic tourism, especially rural-centered leisure tourism, because many city people are unaware of the scenic spots and delicious local cuisine in some rural villages which are not far from the cities. The cost of homestay is cheap, and rural tourism is not crowded. There is fresh air in the countryside to breathe. It is also fun for parents to take their children on road trips in a family car during school holidays. Many people are buying recreational vehicles (RV) for travel around the country. Local governments spend money to build access roads to rural areas and assistance to the homestay tourism industry. Homestay business is popular in the US. China is a vast country with lots of beautiful pristine mountains and rivers in the western regions that should attract local tourists. Domestic tourism can promote sustainable economic development in the countryside and villages. This helps the country's dual circulation policy.

Many university graduates have returned to their home towns in the villages to start online businesses related to tourism and local produce. Across the vast country with scenic natural landscapes of mountains, rivers and forests, homestay business is potentially a huge market to contribute employment and economic activities. Many city residents take their children to famous and popular tourist spots during public holidays such as National Day and New Year. In these popular tourist sites, too many tourists congregate at the same time, causing chaos and congestion. Such holiday travel can turn into a conflict between the holiday makers. Such a chaotic holiday is not worth the time and money spent, thereby spoiling a hard earned holiday for the entire family. It is advisable for the holiday family to stagger their choice of destination to avoid the usual large crowds during the peak seasons. During peak season, prices of hotels and restaurants in popular destinations like Sanya in Hainan Island soar. Some people remark that it is cheaper to have a holiday in Thailand than in Sanya during peak season.

Tourism is a service industry that encompasses many other industries and provides many jobs and opportunities, strengthens local economies, promotes the development of local infrastructure and contributes to the preservation of the natural environment, cultural assets and traditions. There are 6 products in the tourism industry: 1) accommodation industry includes hotels, motels, resorts, bed-and-breakfasts and campsites 2) food and beverage services 3) recreation activities 4) logistics and transportation 5) security and customer service and 6) shopping. To serve foreign tourists, hotels, restaurants, coffee shops, bars, shopping , entertainment venues, transportation and tourism service providers directly employ many employees to meet the needs of tourists. It is estimated that the economic contribution of 40 foreign visitors could support a family of four.

Tourism provides a good window for China to showcase its soft power to the world. It also shows the world the daily life of Chinese people and lets tourists understand Chinese history and culture. From December 1, 2023, China began granting visa-free entry to tourists from Thailand, Germany, France, Italy, the Netherlands, Spain and Malaysia. Later, from March 14 to November 30, 2024, China implemented a unilateral visa-free policy for ordinary passport holders from Switzerland, Ireland, Hungary, Austria, Belgium and Luxembourg. There were 95 million inbound tourists during the first three quarters of 2024, a 78.8% increase year-on-year. China's visa free for foreigners to visit China is a policy that could have been introduced earlier to open up China to the world and to export China's soft images.

China has so many new things to offer foreigners to see the advancement in the country such as the high speed train in which passengers can order foods on the train and get delivered at the next station, free hi-fi available in subways, payment of goods and services by mobile phone without paper money, driverless EV taxi, clean streets without homeless people in the cities, people dancing in the parks at night, absolute safety for people walking in the streets at night, drone delivery at the park for drinks and food ordered online, free use of hi-tech public toilets. China is a country with advanced technology being used in daily life.

22.5 PROBLEMS FACED BY TOURISM INDUSTRY

Compared with other countries with less natural and historical sites attractions, China's tourism revenue from foreign tourists is miniscule. In fact, neither Malaysia nor Singapore has much natural beauty or historical sites, but these two countries attract many tourists from China and other countries. Thailand doesn't have much to offer Chinese tourists beyond beautiful beaches, and yet the country expects 10 million Chinese visitors to Thailand in 2024. The tourism industry in these countries is so prosperous that it is worth taking a leaf from them by the Chinese tourism department. Their governments make sure that tourists are welcome to visit their country and that their travel is safe. The Singapore government protects foreign tourists seriously by taking legal action against a fraudulent merchant who cheated a foreigner by buying a mobile phone without warranty. The merchant was sentenced to jail by the court.

According to reports, the government carried out inspections of service quality at some 5A scenic spots based on feedback of complaints from tourists The inspection results revealed that the following scenic spots shortcomings: entry price fraud, unhygienic environmental sanitation, widespread damaged facilities, lack of proper site maintenance, poor quality of tour guide service, taxis drivers overcharged passengers, disorderly parking lot, dirty toilet facilities. The supervisors of all the mismanaged scenic spots, as well as the local governments were insensitive to and condone some illegal practices. These mismanagement problems have been around for sometime without oversight by the central government. The depth and scope of the problems vary from different scenic spots. To rectify the problems completely throughout the country need detailed investigation and understanding of the root causes. Problems could be related to

- Local government has no interest in managing a scenic spot
- No directives on how to manage the scenic spots without experience or training
- No financial incentives to the local government
- No standard of managing the spots issued by the government
- Insufficient budget to maintain the entire facility properly
- Insufficient staff to manage the facility when the number of visitors vary
- according to seasons
- A huge crowd of visitors during peak periods can overwhelm the local authority and is beyond the control of the staff to maintain rules and orders.
- Some visitors could damage the facility infrastructure and amenities.
- How to manage the parking problems when the parking space is insufficient to accommodate too many buses and cars during peak seasons.

The central government may need to examine whether it is prudent for various local governments to manage the national assets throughout the country and whether there is an urgent need to renovate and upgrade some existing dilapidated facilities . There may be a need to charge an entry fee to sustain the facility management and upkeep. During peak seasons, temporary staff is required to maintain order. Parking spaces may need to be expanded to accommodate more vehicles during peak periods.

The perennial chronic problem of overcrowding at many tourist sites across China during festive seasons and national holidays will have to be addressed to achieve an amicable solution to all visitors in future. One of the plausible methods is to spread the number of visitors to different times of the year. Such a method needs to impose a charge of entering the tourist sites. The amount of entry charge varies according to the peak and lull seasons. To encourage people to visit tourists sites during the lull period, the government can offer free entry to all the tourist sites

22.6 China to Reform Tourism

The government is to establish a ministry of tourism encompassing the management and control of all natural historic and cultural sites and natural sceneries around the country. The tourism industry could be worth more than US$1 trillion in the future due to its vast tourist attractions with 56 diversified ethnic groups and large varieties of local cuisines. China has 5000 years of continuous history that should fascinate many foreign visitors. China has not been promoting tourism in overseas countries in an aggressive way unlike other Asian countries like Japan, Thailand, Malaysia and Singapore.

Opening up China with visa free to foreign visitors from North America, Europe, Middle East, Africa and South America will be a good start. Many well travelled tourists like to explore new adventures in China, considered by some as a mysterious ancient country. Foreign sports enthusiastic snow skiers no need to go to Switzerland to ski as they can go to ski on China well-managed ski slopes that have hosted the winter Olympic in 2008. The Tourism Ministry has a video for each of the scenic spots, cultural and historical sites, with an introduction and history narration. The videos can be downloaded from the China Tourism App. The videos have different foreign language descriptions. These videos help foreigners learn more about and appreciate Chinese culture prior to and during their trip to China.

The Ministry will partner with airlines, hotels, tour guides, logistics, advertising agents, media and transportation to prepare a comprehensive annual plan for tourists of various countries using Deepseek AI. The Ministry studies each country's culture and history, economy, demographic, income levels, travel habits, school and public holidays, their usual holiday destinations during summer and winter. The Ministry will hold travel fairs in each country to promote travel to China, with travel agents offering various travel packages and destinations. The Ministry will advertise on social media in different countries. Many hotels will have TV channels with films featuring travel logs of foreign travelers in China and various festive celebrations of different ethnic minorities. The TV channel also shows the advancements of science and technology in China and in people's daily life.

According to a report, an estimated 1.4 billion tourists traveled internationally in 2024 generating a record $1.9 trillion in revenue. China's inbound tourism receipts contribution to the national GDP is minuscule . If China wants to earn the same amount of US tourism receipt per capita of $519, China needs to have a total tourism receipt of USS$727b from 182m tourists. With passage of time when China becomes a favorite country for foreigners to visit and when the world becomes richer, China will soon achieve an increase of tourists to reach 259 million from all over the world. By then the total receipts would reach US$1 trillion. There will be many return tourists as China is too huge a country to see in one trip. A booming tourism industry will help Comac sell more airplanes to both domestic and foreign airlines. Tourism helps develop the Chinese economy. Here is the list of country's foreign tourism receipts per capita: Singapore $3170, Spain $1916, France $1064. Italy $964, USA $519, China $17.7

There are many scenic spots such as waterfalls in China that have not been discovered or developed yet. In order to cater to the growing number of domestic and foreign tourists and to reduce the pressure on existing tourist attractions, these virgin scenic spots should be developed. There are many tourist attractions in the coastal southern provinces, where there are beautiful seaside beaches for swimming and water sports for summer and winter vacationers from northern China and foreign countries. China has many hidden scenic spots in Xinjiang where majestic mountains and clear water rivers and lakes have not been publicized on the tourist maps as well as archaeological sites developed as tourist attractions. With more new tourist attractions developed together with new infrastructure, the pressure on existing famous tourist attractions will be relieved and the quality of tourism will be improved.

From Ancient to Futuristic World

Michael Loong